A TREATISE

ON

SPHERICAL TRIGONOMETRY

MB

A TREATISE

ON

SPHERICAL TRIGONOMETRY

And Its Application To

Geodesy And Astronomy

With

Numerous Examples

By

John Casey, LL.D., F.R.S.,
Royal University Of Ireland

MB

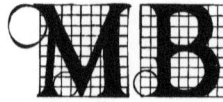

PREFACE.

THE present Manual is intended as a Sequel to the Author's Treatise on Plane Trigonometry, and is written on the same plan. An examination of the Table of Contents, or of the Index, will show the scope of the work. It will be seen that, though moderate in size, it contains a large amount of matter, much of which is original.

The sources from which I have obtained information are indicated in the text. The principal are CRELLE's *Journal* "für die reine und angewandte Mathematik," Berlin, and *Nouvelles Annales de Mathématiques*, Paris.

The examples, which are very numerous (over five hundred) and carefully selected, illustrate every part of the subject. Among them will be found some of the most elegant Theorems in Spherical Geometry and Trigonometry.

In the preparation and arrangement of every part of the work I have received invaluable assistance from Professor Neuberg, of the University of Liège. For this, as well as for similar assistance previously given in the editing of my *Plane Trigonometry*, I beg to return that gentleman my most grateful acknowledgments and best thanks.

JOHN CASEY.

86, South Circular Road, Dublin.
March 25, 1889.

CONTENTS.

CHAPTER I.

SPHERICAL GEOMETRY.

CHAPTER II.

FORMULAE CONNECTING THE SIDES AND ANGLES OF A SPHERICAL TRIANGLE.

CHAPTER III.

SOLUTION OF SPHERICAL TRIANGLES.

CHAPTER IV.

VARIOUS APPLICATIONS.

CHAPTER V.

SPHERICAL EXCESS.

CHAPTER VI.

SMALL CIRCLES ON THE SPHERE.

CHAPTER VII.

INVERSIONS.

CHAPTER VIII.

POLYHEDRA.

CHAPTER IX.

APPLICATIONS OF SPHERICAL TRIGONOMETRY.

ERRATA.

Page   3, last line, omit "the".

  ,,   9, line   6, *for BD = read BD +.*

  ,,   22, ,,   10, *for* cos a cos b cos c, *read* cos a − cos b cos c.

  ,,   72, ,,   13, insert a comma after $(O − ABC)$.

  ,, 112, ,,   11, *for x, y, z, read* sines of x, y, z.

  ,, 131, ,,   Exercise 4, *for* three times, *read* one-third.

SPHERICAL TRIGONOMETRY.

CHAPTER I.

SPHERICAL GEOMETRY.

Section I.—Preliminary Propositions and Definitions.

1. Def. I.—*A sphere is the surface generated by the revolution of a semicircle about its diameter, which remains fixed.*

The term *sphere* is used in a two-fold signification—1°. As denoting the surface. 2°. The solid bounded by the surface. These correspond to the two-fold signification of the word *circle* in plane Geometry, namely, the *circumference*, and the area included within it.

Def. II.—*The centre of the generating semicircle is called the* CENTRE *of the sphere.*

Def. III.—*A* RADIUS *of the sphere is any right line drawn from the centre to a point in the surface.*

Def. IV.—*A* DIAMETER *of a sphere is any right line drawn through the centre, and terminated both ways by the surface.*

From the definition of a spherical surface it follows at once—1°. That every point in it is equally distant from the centre of the generating semicircle. 2°. That any point *P* in space is *outside*, *on*, or *inside* the surface, according as its distance from the centre is *greater* than, *equal to*, or *less* than the radius. 3°. That spheres having equal radii are equal.

2. *Every section of a sphere made by a plane is a circle.*

1°. If the plane passes through the centre, such as *ABC*, the proposition is evident, since every point in the surface is equally distant from the centre.

2°. When the plane does not pass through the centre, such as *DEF*. Let *O* be the centre. From *O* let fall a perpendicular *OI* on *DEF* (Euc. XI. xi.). Take any point *F* in the section *DEF*. Join *OF*, *IF*. Then, since *OI* is normal to the plane, the angle *OIF* is right; therefore $IF^2 = OF^2 - OI^2$; but *OF* is constant, being the radius of the sphere. Hence *IF* is constant, and therefore the section *DEF* is a circle, whose centre is *I* and radius *IF*.

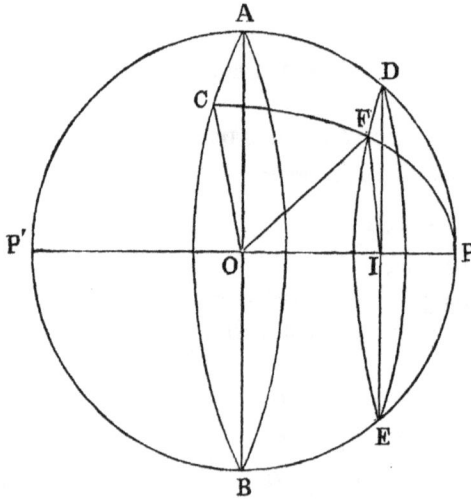

Fig. 1.

Cor. 1.—If *R* be the radius of the sphere, *r* the radius of the section, *d* the distance of the plane of section from the centre of the sphere,

$$r^2 = R^2 - d^2. \tag{1}$$

Cor. 2.—If $R = d$, $r = 0$. Hence the section will reduce to a point, and the plane will touch the sphere.

Cor. 3.—Two circles, whose planes are equally distant from the centre, are equal.

DEF. V.—*The two points P, P' in which the diameter perpendicular to the plane of the circle DEF meets the sphere are called its* POLES.

From this definition it follows—1°. That all circles whose planes are parallel have the same poles. 2°. That the centre of any circle, its poles, and the centre of the sphere, are collinear.

DEF. VI.—*A circle of the sphere whose plane passes through the centre is called a* GREAT CIRCLE, *and a circle whose plane does not pass through the centre is called a* SMALL CIRCLE. Thus, on the earth, the meridians, the equator, the ecliptic, are *great circles*, and the parallels of latitude are *small circles*.

3. *The curve of intersection of two spheres is a circle.*

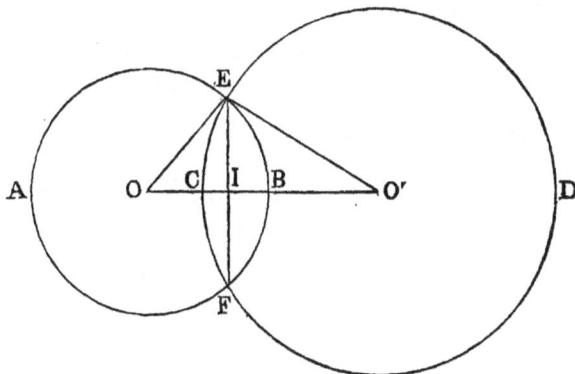

Fig. 2.

DEM.—Let any plane passing through the centres O, O' of both spheres cut them in the circles *AEBF, CEDF*. Join *EF, OO'*, and produce *OO'* to meet the circles in *A, D*. Now *OO'* bisects *EF* perpendicularly in *I*, and it is evident when the semicircles *AEB, CED* revolve round the line *OO'* to describe the spheres, that the point *E* will describe a circle, having ~~the~~ *I* for its centre.

4. *Either pole of a circle (great or small) on the sphere is equally distant from every point in its circumference.*

For (see fig., prop. II.), join PF. We have $PF^2 = PI^2 + IF^2$; but IF^2 is constant $= R^2 - d^2$, and $PI^2 = (R - d)^2$. Hence PF is constant.

Cor. 1.— $\qquad\qquad PF^2 = 2R\,(R - d).\qquad\qquad (2)$

Cor. 2.— $\qquad\qquad P'F^2 = 2R\,(R + d).\qquad\qquad (3)$

DEF. VII.—*A great circle passing through the poles of another circle (great or small) is called a secondary to that circle.*

DEF. VIII.—*The spherical radius of a small circle is the arc of a secondary, intercepted between any point in the circumference and the nearest pole.* Thus the spherical radius of the small circle DEF (see fig., prop. II.) is the arc PD.

Cor. 1.—If OA be perpendicular to OP, the point A will describe a great circle.

Cor. 2.—The spherical radius of a great circle is a quadrant.

This is evident; since P, P' are the poles of the great circle ABC, and AP, CP are quadrants.

5. *Only one great circle can be drawn through two points on the surface of the sphere, unless they are diametrically opposite.*

For only one plane can be drawn through the centre and the two points, unless they are collinear.

Cor. 1.—If two points A, C be each 90° distant from a third point P, P is the pole of the great circle, determined by the points A, C.

If O be the centre, the line OP is perpendicular to the lines OA, OC, and therefore it is normal to their planes. Hence the line PP' is the axis of the great circle in which the plane OAC cuts the sphere, and P, P' are its poles.

Cor. 2.—If the planes of two great circles be at right angles to each other, their axes are perpendicular, and each passes through the poles of the other.

6. *The locus of all the points of a sphere which are equidistant from two fixed points A, B of the sphere, is the great circle, which is perpendicular at its middle point to the arc of the great circle AB.*

Dem.—Let *C* be the middle point of the chord *AB*. At *C* erect a plane *P*, perpendicular to the chord *AB*. *P* passes through the centre of the sphere, and it is the locus of points equally distant from *A* and *B*; therefore the points of the sphere, where *P* intersects it, are the only points on it which are equidistant from the points *A*, *B*. Hence the proposition is proved.

7. *Any two great circles of the sphere bisect each other.*

Dem.—Let *ABCD*, *AECF* be the great circles; then (Def. vi.) the plane of each passes through the centre of the sphere. Hence the common section *AC* of these planes passes

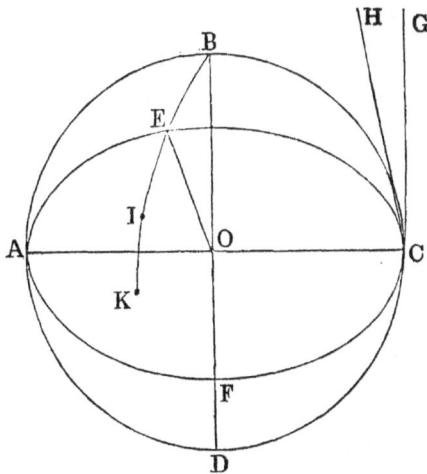

Fig. 3.

through the centre; but the common section of two planes is a right line (Euc. XI., iii.) Hence *AC* is a diameter of the sphere; therefore *ABC*, *AEC* are semicircles, and the proposition is proved.

Def. IX.—*When two arcs of circles intersect, the angle of the tangents at their points of intersection is called the angle of the arcs.*

8. *The angle of intersection of two great circles is equal to the inclination of their planes.*

Dem.—Let CG, CH be tangents to the semicircles ABC, AEC; then (Euc. XI., Def. ix.), since each is perpendicular to OC, the angle between them is equal to the angle of inclination of the planes of the great circles; but (Def. ix.) the angle between CG, CH is the angle of intersection of the great circles. *Therefore the angle between two great circles is equal to the inclination of their planes.*

Cor. 1.—If CB, CE be quadrants, OB, OE are at right angles to OC, and the angle BOE is equal to the angle of inclination of the planes of the great circles. Hence the spherical angle BCE is equal to the angle BOE; but BOE is measured by the arc BE. *Hence the spherical angle contained by any two great circles ABC, AEC is measured by the arc of a great circle intercepted between them, and having the point C for its pole.*

Cor. 2.—The spherical angle BAE is equal to the angle BCE.

Cor. 3.—The angle of intersection of two great circles is measured by the arc between their poles.

For, if BE be produced, since the plane BOE is perpendicular to OC, BE will pass through the poles of ABC, AEC. Let these be I, K, respectively; then, evidently, the arcs IB, KE are quadrants. Hence $IK = BE$; but BE (*Cor.* 1) is the measure of the spherical angle BCE. Hence IK is equal to the measure of the spherical angle.

9. *To find the radius of a solid sphere.*

Sol.—From any point of the spherical surface as pole, with any arbitrary opening of the compass describe a circle ABC;

in the circumference of this circle take any three arbitrary points *A, B, C*, and with the compass transfer the three recti-linear distances *AB, BC, CA*, and construct a triangle *abc* on paper, having its sides *ab, bc, ca* respectively equal to *AB, BC, CA*. Find *i*, the circumcentre of the triangle *abc*. Join *ia*.

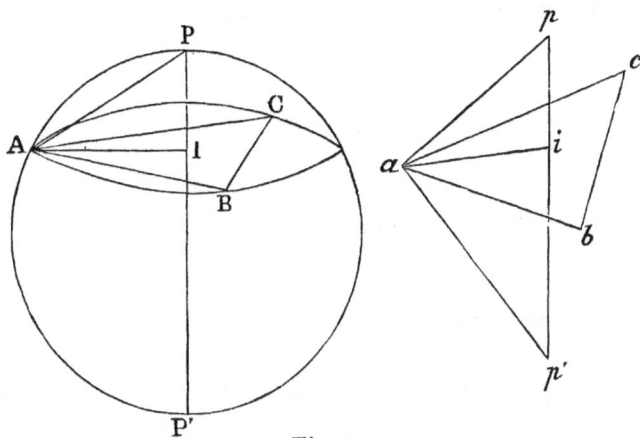

Fig. 4.

Erect *ip* perpendicular to *ia*, and inflect from *a* to *ip* the dis-tance *ap* equal to the opening of the compass with which the circle *ABC* was described on the sphere. Erect *ap'* at right angles to *ap*, and produce *ip* to meet it in *p'*; then *pp'* is equal to the diameter of the solid sphere. For, from the construction, it is evident, if we join *AP'*, that *pp'* is equal to *PP'*.

10. *Analogy between the geometry of the sphere and the plane.*

In order to understand the analogy between plane and sphe-rical geometry, it is necessary to observe that to right lines on the plane correspond on the sphere great circles, and to circles on the plane correspond circles on the sphere, which may be either great or small.

On a solid sphere can in general be resolved problems analo-gous to those on the plane, the instrument employed being the compass. Thus (see fig., § 2), if we place one of the points of the compass in *P*, we can, with an opening equal to the chord *PD*, describe the circle *DEF*. To describe a great circle, it is necessary that the opening of the compass should be equal to the chord of a quadrant. We can also, by the compass, divide an angle into 2, 4, 8, &c., equal parts, erect an arc of a great circle perpendicular to another, make a spherical angle equal to a given spherical angle, describe a circle touching three given circles, &c.

Exercises.—I.

1. A great circle passing through the poles of two others cuts each at right angles, and their points of intersection are its poles.

2–5. Solve the following problems with the compass :—

1°. Describe a great circle through two given points of the sphere.

2°. Through a given point of the sphere draw an arc of a great circle perpendicular to a given great circle.

3°. Make, at a given point of a given great circle, an angle equal to a given angle on the same sphere.

4°. Through a given point not on a given great circle draw a great circle making a given angle with it.

6. The loci of the poles of great circles, making a given angle a with a given great circle, consist of two small circles, having the same poles as the given circle.

7. The tangents at a given point A of the sphere to all circles (great or small) passing through A lie in the plane through A, perpendicular to the radius of the sphere drawn to that point.

8. If a tangent line to a sphere passes through a given point, the locus of the point of contact is a small circle.

9. If tangent lines to a sphere be parallel to a given line, the locus of the points of contact is a great circle.

10. The arc of a great circle, perpendicular to the spherical radius of a small circle at its extremity, touches the small circle.

11. Draw a great circle, touching two small circles.

12. Draw a great circle through a given point, touching a given small circle.

13. If a variable sphere touch three planes, the locus of its centre is a right line.

14. Describe a sphere, passing through four points which are not co-planar.

15. If A, B, C, D be four points on a great circle, prove that

$$\sin BC . \sin AD + \sin CA . \sin BD + \sin AB . \sin CD = 0.$$

16. In the same case, prove that

$$\sin BC . \cos AD + \sin CA \cos BD + \sin AB \cos CD = 0.$$

Section II.—Spherical Triangles.

11. Def. X.—*The figure formed by the shorter arcs joining three points on the surface of a sphere, no two of which are diametrically opposite, is called a* spherical triangle.

Two points on the surface of a sphere can be joined by two distinct arcs, which together make a great circle. Hence, when the points are not diametrically opposite, these arcs are unequal, and it follows from the definition that each side of a spherical triangle is less than a semicircle.

If ABC be the triangle, O the centre of the sphere, the planes OAB, OBC, OCA form a solid angle $O - ABC$

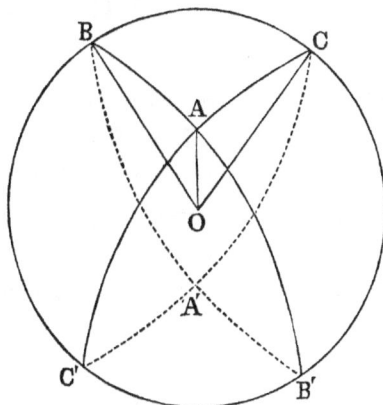

Fig. 5.

(Euc. XI., Def. iii.), whose face angles AOB, BOC, COA are measured by the sides of the spherical triangle ABC, and whose dihedral angles (Euc. XI., Def. II.) are equal to the angles of the spherical triangle (§ 8). There is then a correspondence between the spherical triangle ABC and the solid angle $O - ABC$: every property of one gives a property of the other.

DEF. XI.—*The portion of a sphere comprised between two halves of great circles is called a* LUNE.

Three great circles intersect in six points A, A'; B, B'; C, C'. These are two by two diametrically opposite, and divide the sphere into eight triangles.

DEF. XII.—*Two triangles BCA, $B'CA$, which have a common side CA, and whose other sides belong to the same great circles, are called* COLUNAR TRIANGLES. The triangle ABC has three colunar triangles, viz. $A'BC$, $B'CA$, $C'AB$.

DEF. XIII.—*Two triangles ABC, $A'B'C'$, whose corresponding vertices are diametrically opposite, are called* ANTIPODAL TRIANGLES.

12. *Any two sides of a spherical triangle are together greater than the third, and the sum of the three sides is less than a great circle.*

DEM.—Let ABC be the spherical triangle (see fig., § 11), O the centre of the sphere. Join OA, OB, OC; then (Euc. XI., xx.) any two of the plane angles forming the trihedral angle $O - ABC$ are together greater than the third; but the arcs AB, BC, CA are the measures of the plane angles AOB, BOC, COA. *Hence the sum of any two of the arcs AB, BC, CA is greater than the third.*

Again, the sum of the three plane angles AOB, BOC, COA is (Euc. XI., xxi.) less than four right angles. *Hence the sum of the three arcs AB, BC, CA is less than a great circle.*

In the same manner it follows *that the sum of the sides of any convex spherical polygon is less than a great circle.*

13. Since every great circle has two poles, it will be necessary to make some convention in order to distinguish them. For this purpose we employ, as in so many other cases, the terms *positive* and *negative*. Thus, if *BC* be an arc of a great circle, its *positive* pole will be that round which the rotation from *B* to *C* will be from left to right; that is, in the same direction in which the hands of a watch move, and the other will be the *negative* pole.

For example, if *B*, *C* be points on the equator, and *C* west of *B*, the *north pole* will be the positive pole of *BC*, and the *south* its negative pole.

DEF. XIV.—*The spherical triangle, whose angular points are the positive poles of the sides of a triangle ABC is called the* POLAR TRIANGLE *of ABC.*

14. *If two spherical triangles ABC, A′B′C′ be such that the latter is the polar triangle of the former, the former is the polar triangle of the latter.*

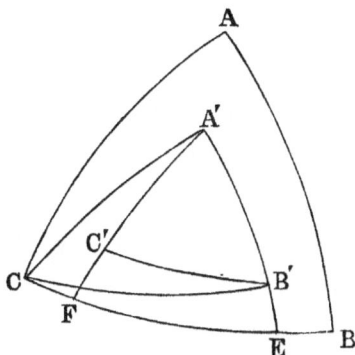

Fig. 6.

DEM.—Join *A′C*, *B′C* by arcs of great circles; then, because *A′* is the pole of *BC*, *A′C* is a quadrant. Similarly, *B′C* is a quadrant. Hence, since the arcs *CA′*, *CB′* are quadrants, *C* is the pole of *A′B′*, and it is evidently the positive pole. Hence the proposition is proved.

15. *The sides of either of two polar triangles are the supplements of the angles of the other.*

DEM.—Produce the arcs $A'B'$, $A'C'$ if necessary to meet BC in E and F. Now, since C is the pole of $A'B'$, the arc EC is 90°. In like manner, the arc BF is 90°. Hence $EC + BF$ or $BC + EF$ is 180°; but EF is (Art. 8, *Cor.* 1) the measure of the spherical angle $B'A'C'$. Hence the side BC is the supplement of the angle $B'A'C'$, and similarly for the other sides and angles.

Scholium.—On account of the property proved in this proposition, *polar triangles* are also called *supplemental triangles.*

Cor.—If we denote the angles of the triangle ABC by A, B, C; their opposite sides by a, b, c; and the corresponding elements in the polar triangle by the same letters accented, we have

$$a' = 180° - A, \qquad A' = 180° - a. \qquad (4)$$

$$b' = 180° - B, \qquad B' = 180° - b. \qquad (5)$$

$$c' = 180° - C, \qquad C' = 180° - c. \qquad (6)$$

16. *In every spherical triangle—*1°. *The sum of two angles is less than the third increased by* 180°. 2°. *The sum of the three angles is greater than two, and less than six right angles.*

DEM.—1°. From the equations (4)–(6) we get

$$(a' + b' - c') = 180° + C - (A + B);$$

but $a' + b'$ is greater than c', therefore $180° + C$ is greater than $A + B$.

2°. From the same equations, we have

$$A + B + C = 540° - (a' + b' + c').$$

Hence $A + B + C$ is less than 540°; that is, six right angles. Again (from § 12), $a' + b' + c'$ is less than 360°. Hence $A + B + C$ is greater than 180°; that is, two right angles.

Cor. 1.—If any side of a spherical triangle ABC be produced, the exterior angle is less than the sum of the two
interior non-adjacent angles.

Cor. 2.—If all the sides of a convex spherical polygon be
produced, the sum of the exterior angles is less than four
right angles.

DEF. XV.—*The amount by which the sum of the three angles of
a spherical triangle exceeds two right angles is called the spherical
excess.* *We shall denote it by* $2E$.

Denoting the spherical excess by $2E$ instead of E has the same advantage
as putting $2s$ for the perimeter of a triangle instead of s, viz., it avoids
fractions, and makes certain formulae containing angles symmetrical with
the corresponding ones containing sides.

Cor. 3.—Any angle of a spherical triangle is greater than E.
This is merely another statement of 1°, *supra*.

17. *Two antipodal triangles* ABC, $A'B'C'$ *are equal in area.*

DEM.—Two antipodal triangles have evidently equal sides,
but are not superposable, except when each is isosceles, because

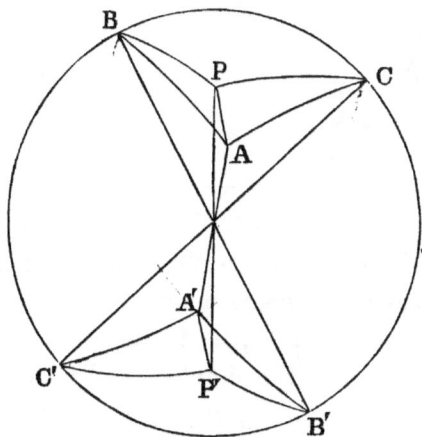

Fig. 7.

their elements are arranged in inverse order. To prove that in
the general case the areas are equal. Let P be the pole of the

small circle, passing through A, B, C, and P' the pole of the circle through $A'B'C'$; then evidently P' is diametrically oppo-site to P, and the pairs of triangles PAB, $P'B'A'$; PBC, $P'C'B'$; PCA, $P'A'C'$, being antipodal and isosceles, are superposable. Hence the area of ABC is equal to the area of $A'B'C'$.

18. *Two spherical triangles on the same sphere have all their corresponding elements equal—1°. When two sides and the con-tained angle of one are respectively equal to two sides and the con-tained angle of the other. 2°. When the side and the adjacent angles of one are equal to a side and the adjacent angles of the other. 3°. When the three sides of one are equal to the three sides of the other. 4°. When the three angles of one are equal to the three angles of the other.*

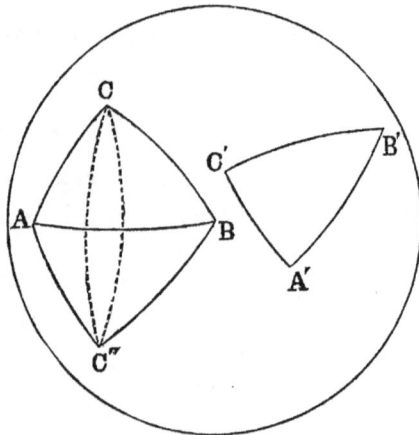

Fig. 8.

Cases 1°, 2°, 3° correspond to Euc. Book I., Props. IV., VIII., XXVI. Case 4° has no analogue in *Plane Geometry*. It will be sufficient to prove 1° and 3°, as 2° and 4° are inferred from them by the properties of the supplemental triangle.

DEM. 1°.—$A = A'$, $AB = A'B'$; $AC = A'C'$. If these ele-ments are arranged in the same order, the demonstration follows by superposition, as in *Plane Geometry*. If they are disposed in an inverse order, such as $A'B'C'$, ABC'', we can superpose either of them on the antipodal triangle of the other.

3°. If the arc $A'B'$ be applied to AB, the point C' will be one of the points of intersection of the arcs of two small circles, described from A and B as poles, and passing through the point C: these arcs will intersect in two points C, C''', placed on opposite sides of AB; then, if the elements are disposed in the same order in both triangles, C' will coincide with C. If in a different order, the triangle $A'B'C'$ can be superposed on the antipodal triangle of ABC, and in each case we have the corresponding angles equal each to each.

19. *If two sides AB, AC of a spherical triangle be equal—* 1°. *The angles B, C are equal.* 2°. *The median AD, which bisects BC, bisects the angle A.*

Dem.—The arc AD divides the triangle ABC into two triangles, which are symmetrically equal (18, 3°).

Cor.—If two angles B, C are equal, the opposite sides AB, AC, are equal. For the polar triangle of ABC has two sides equal. Hence, &c.

20. *To find the area of a lune.*

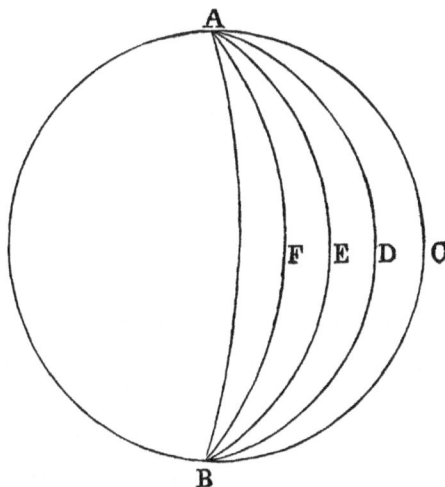

Fig. 9.

Let $ACBD$, $AEBF$ be two lunes having equal angles at A; then, by superposition, it is evident that these lunes are equal.

Hence, by a process similar to that employed in Euc. VI., I. and XXXIII., it may be proved that lunes are proportional to their angles. Therefore a lune : the whole spherical surface : : angle of lune : 2π. Now if r denote the radius of the sphere, its surface is $4\pi r^2$ (Euc., App. 7). Hence, if A denotes the angle of the lune, its area is $2Ar^2$. (7)

21. Girard's Theorem.—

The area of a spherical triangle = $2Er^2$ (Def. xv.).

Dem.—Produce the base AB round the sphere, and produce BC, AC to meet it in E and D; also produce CB, CA through B and A to meet again in F. Then the spherical triangle BAF is antipodal to the triangle EDC, and therefore (Art. 17) equal in area to it. Hence the lune C is equal to the sum of

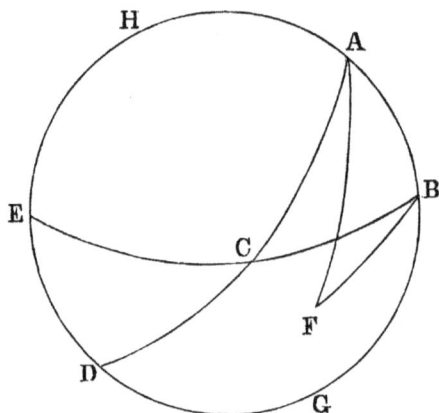

Fig. 10.

the two triangles ABC, CED; also the lune $A = $ to the sum of the triangles ABC, BCD, and the lune $B = $ to the sum of ABC, CEA. Hence the sum of the three lunes is equal to twice the area of the spherical triangle ABC, together with the area of the hemisphere $C = ABGDEH$. Hence, if Δ denote the area of the triangle ABC, we have

$$2Ar^2 + 2Br^2 + 2Cr^2 = 2\pi r^2 + 2\Delta;$$
$$\therefore \quad \Delta = (A + B + C - \pi)\,r^2 = 2Er^2. \quad (8)$$

This demonstration is taken from the works of WALLIS, Vol. II., p. 875. The theorem is due to ALBERT GIRARD, a Flemish Mathematician of the 17th century. In 1787, more than 150 years after its discovery, an important application of it was made by General Roy in correcting the spherical angles, observed in the Trigonometrical Survey of Britain, *Phil. Trans.*, Vol. VIII., p. 163, year 1790. See also *Mem. Acad.*, Paris, 1787, p. 358, and *Mem. Inst.*, Vol. VI., p. 511.

Cor. 1.—The area of a great circle : area of the spherical triangle : : $\pi : 2E$. (9)

Cor. 2.—If Σ denote the sum of the angles of a spherical polygon of n sides, its area is

$$\{\Sigma + (2 - n)\pi\} \, r^2. \tag{10}$$

EXERCISES.—II.

1. If a triangle coincides with its supplemental triangle, prove that all its sides are quadrants and all its angles right.

2. The sum of two opposite angles of a spherical quadrilateral inscribed in a small circle is equal to the sum of the two others, and each sum is greater than two right angles.

3. The spherical excess of a spherical triangle is equal to the circumference of a great circle diminished by the perimeter of the supplemental triangle.

4. The sum of two opposite sides of a spherical quadrilateral, circumscribed to a small circle, is equal to the sum of the remaining sides.

5. If A, B, C, D be four concyclic points on a sphere, prove that

$$\sin \tfrac{1}{2} BC . \sin \tfrac{1}{2} AD + \sin \tfrac{1}{2} CA . \sin \tfrac{1}{2} BD + \sin \tfrac{1}{2} AB . \sin \tfrac{1}{2} CD = 0. \tag{11}$$

This follows from Ptolemy's theorem, since chord $BC = 2 \sin \tfrac{1}{2} BC$, &c.

6. In the same case, if

$$AB = a, \quad BC = b, \quad CD = c, \quad DA = d, \quad AC = e, \quad BD = f;$$

prove that

$$\frac{\sin \tfrac{1}{2} e}{\sin \tfrac{1}{2} f} = \frac{\sin \tfrac{1}{2} a . \sin \tfrac{1}{2} d + \sin \tfrac{1}{2} b . \sin \tfrac{1}{2} c}{\sin \tfrac{1}{2} a . \sin \tfrac{1}{2} b + \sin \tfrac{1}{2} c . \sin \tfrac{1}{2} d}. \tag{12}$$

DEF. XVI.—*A spherical triangle ABC is said to be diametrical when its circumcentre is the middle point D of one of its sides AB.   This side is called the* DIAMETRICAL SIDE.

7. In a diametrical triangle, the angle opposite the diametrical side is equal to the sum of the two remaining angles, and is greater than a right angle.

8. Two of the colunar triangles of a diametrical triangle are also diametrical triangles, and the spherical excess of the third colunar triangle is equal to two right angles.

9. If the opposite sides of a spherical quadrilateral be equal, the diagonals bisect each other, and the opposite angles are equal.

10. If in a spherical quadrilateral $ABCD$ the angle $A = C$ and $B = D$; then the side $AB = CD$, and $BC = AD$.

Produce the sides AB, CD to meet in E and F; then triangles EBC, FAD have the three angles of one respectively equal to the three angles of the other.

DEF. XVII.—*If the diagonals of a spherical quadrilateral bisect each other, it is called a* SPHERICAL PARALLELOGRAM.

11. If the four sides of a spherical quadrilateral be equal, the diagonals are perpendicular to each other, and they bisect its angles. *Such a figure is called a* SPHERICAL LOZENGE.

12. If the four angles of a spherical quadrilateral be equal, the diagonals are equal.

13. In two supplemental triangles ABC, $A'B'C'$, the arcs AA', BB', CC' are perpendiculars to the corresponding sides of the two triangles, and the corresponding altitudes of the two triangles are supplemental.

14. The poles of the small circle inscribed in a spherical triangle are also the poles of the small circle circumscribed to its supplemental triangle, and the spherical radii of both circles are complementary.

15. If two small circles on a sphere touch each other, the angle between their planes is equal to the sum or the difference of their spherical radii.

16. The angle of intersection of a great circle and a small circle is greater than the inclination of their planes.

17. The length of a degree on a parallel of latitude is equal to the length of a degree of the equator multiplied by cos lat.

For if r be the radius of the equator, and r' the radius of the parallel, then, degree on parallel divided by degree on equator $= r'/r = \cos$ lat.

CHAPTER II.

FORMULAE CONNECTING THE SIDES AND ANGLES OF A SPHERICAL TRIANGLE.

22. A spherical triangle has six elements, namely, the three sides a, b, c, and the three angles A, B, C respectively opposite to them. The triangle is completely determined when any three of the six elements are given, as there exist relations between the given and the sought parts by means of which the latter may be found. The object of this chapter is to establish these relations. Our formulae will be divided into three classes as follows:—The first class includes all formulae into which enter four elements of the triangle. The second those which contain five elements, and the third class the formulae into which enter all six. The formulae which we are going to investigate apply equally to "trihedral angles." The sides of the spherical triangle correspond to the plane angles, forming the trihedral, and the angles of the spherical triangle to the dihedral angles of the trihedral.

SECTION I.—FIRST CLASS.

23. There are four Cases of the First Class:—

    I. Three Sides and an Angle.

    II. Two Sides and the Angle opposite to one of them.

    III. Two Sides and two Angles, one of which is contained by the sides.

    IV. Three Angles and a Side.

Case I.—Three Sides and an Angle.

24. Let ABC be a spherical triangle, O the centre of the sphere. Join OA, OB, OC. From any point D in OA draw in the planes AOB, AOC, respectively, the lines DE, DF, at right angles to OA. Then (Euc. XI., Def. IX.), the angle EDF is the inclination of the planes to each other, and therefore (§ 8) is equal to the spherical angle A. Join EF; then, from the plane triangles EOF, EDF, we have

$$EF^2 = OE^2 + OF^2 - 2OE \cdot OF \cos EOF.$$

$$EF^2 = DE^2 + DF^2 - 2DE \cdot DF \cos EDF.$$

Hence $2OE \cdot OF \cdot \cos EOF = 2OD^2 + 2DE \cdot DF \cdot \cos A$;

$$\therefore \quad \cos EOF = \frac{OD}{OF} \cdot \frac{OD}{OE} + \frac{DF}{OF} \cdot \frac{DE}{OE} \cdot \cos A,$$

or, $\cos a = \cos b \cos c + \sin b \sin c \cos A.$ (13)

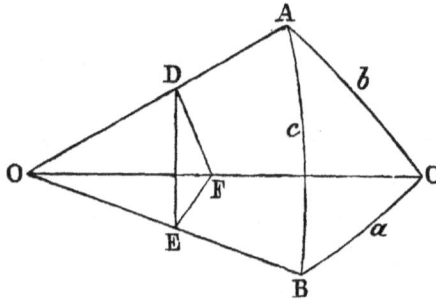

Fig. 11.

This is the fundamental formula of Spherical Trigonometry. By interchanging letters we get

$$\cos b = \cos c \cos a + \sin c \sin a \cos B. \qquad (14)$$

$$\cos c = \cos a \cos b + \sin a \sin b \cos C. \qquad (15)$$

25. The formula (13) has been proved only for the case in which the arcs b, c are less than quadrants, to show that they

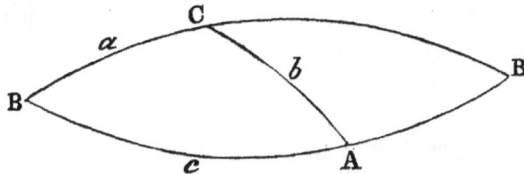

Fig. 12.

hold when one of them, c, is greater than a quadrant.   Produce

BA, BC to meet in B'. Then, from the triangle $B'AC$, in which the sides $B'A$, AC are less than quadrants, we have

$$\cos B'C = \cos B'A \cos AC + \sin B'A \sin AC \cos B'AC,$$

or $\quad \cos (\pi - a) = \cos (\pi - c) \cos b + \sin (\pi - c) \sin b \cos (\pi - A).$

Hence $\qquad \cos a = \cos b \cos c + \sin b \sin c \cos A.$

If both b and c be greater than quadrants, produce AB, AC to meet in A', and the proposition will evidently hold for the triangle $A'BC$, and therefore for ABC.

26. By subtracting equation (13) from the identity

$$\cos (b - c) = \cos b \cos c + \sin b \sin c,$$

we get $\qquad \cos (b - c) - \cos a = \sin b \sin c (1 - \cos A).$

Hence, putting $a + b + c = 2s$, we get

$$\sin \tfrac{1}{2} A = \sqrt{\frac{\sin (s - b) \sin (s - c)}{\sin b \sin c}}. \tag{16}$$

Similarly, $\quad \sin \tfrac{1}{2} B = \sqrt{\dfrac{\sin (s - c) \sin (s - a)}{\sin c \sin a}}, \tag{17}$

and $\qquad \sin \tfrac{1}{2} C = \sqrt{\dfrac{\sin (s - a) \sin (s - b)}{\sin a \sin b}}. \tag{18}$

27. By subtracting the identity

$$\cos (b + c) = \cos b \cos c - \sin b \sin c$$

from (13), we get

$$\cos a - \cos (b + c) = \sin b \sin c (1 + \cos A).$$

Hence $\qquad \cos \tfrac{1}{2} A = \sqrt{\dfrac{\sin s \sin (s - a)}{\sin b \sin c}}. \tag{19}$

Similarly, $\quad \cos \tfrac{1}{2} B = \sqrt{\dfrac{\sin s \sin (s - b)}{\sin c \sin a}}, \tag{20}$

and $\qquad \cos \tfrac{1}{2} C = \sqrt{\dfrac{\sin s \sin (s - c)}{\sin a \sin b}}. \tag{21}$

The radicals in the formulae (16)–(21) have the positive sign; for $\tfrac{1}{2} A$, $\tfrac{1}{2} B$, $\tfrac{1}{2} C$ are each less than 90°.

28. From (16) and (19), we get

$$\tan \tfrac{1}{2} A = \sqrt{\frac{\sin (s - b) \sin (s - c)}{\sin s \sin (s - a)}}. \tag{22}$$

In like manner,

$$\tan \tfrac{1}{2} B = \sqrt{\frac{\sin (s - c) \sin (s - a)}{\sin s \sin (s - b)}}, \tag{23}$$

and

$$\tan \tfrac{1}{2} C = \sqrt{\frac{\sin (s - a) \sin (s - b)}{\sin s \sin (s - c)}}. \tag{24}$$

Cor.—

$$\tan \tfrac{1}{2} A \tan \tfrac{1}{2} B = \frac{\sin (s - c)}{\sin s}. \tag{25}$$

$$\tan \tfrac{1}{2} B \tan \tfrac{1}{2} C = \frac{\sin (s - a)}{\sin s}. \tag{26}$$

$$\tan \tfrac{1}{2} C \tan \tfrac{1}{2} A = \frac{\sin (s - b)}{\sin s}. \tag{27}$$

Exercises.—III.

1. Prove $\sin^2 A = \dfrac{1 - \cos^2 a - \cos^2 b - \cos^2 c + 2 \cos a \cos b \cos c}{\sin^2 b \sin^2 c}$. (28)

Make use of $\cos A = \dfrac{\cos a - \cos b \cos c}{\sin b \sin c}$.

2. ,, $\cos c = \cos (a + b) \sin^2 \tfrac{1}{2} C + \cos (a - b) \cos^2 \tfrac{1}{2} C.$ (29)

3. ,, $\cos^2 \tfrac{1}{2} c = \cos^2 \tfrac{1}{2} (a + b) \sin^2 \tfrac{1}{2} C + \cos^2 \tfrac{1}{2} (a - b) \cos^2 \tfrac{1}{2} C.$ (30)

4. ,, $\sin^2 \tfrac{1}{2} C = \sin^2 \tfrac{1}{2} (a + b) \sin^2 \tfrac{1}{2} C + \sin^2 \tfrac{1}{2} (a - b) \cos^2 \tfrac{1}{2} C.$ (31)

5. ,, $\sin \tfrac{1}{2} A \sin \tfrac{1}{2} B \sin \tfrac{1}{2} C = \dfrac{n^2}{\sin s \sin a \sin b \sin c},$ (32)

where $n = \sqrt{\sin s \sin (s - a) \sin (s - b) \sin (s - c)}.$ (33)

The function n is so important in spherical trigonometry that it is convenient to have a definite name for it. Prof. Staudt*, of the University of Erlangen, calls $2n$ the sine of the trihedral angle $O - ABC$ (see fig., § 24). Neuberg suggests two other names—1°. The *First Staudtian* of the triangle.

---

* Crelle's *Journal*, Band. xxiv., s. 255.

$2°$. The *Norm of the sides* of the triangle; and for the function N (see § 33) he also suggests the names *Second Staudtian*, or the *Norm of the angles of the triangle.*

6. Prove $\quad \cos \frac{1}{2} A \cos \frac{1}{2} B \cos \frac{1}{2} C = \dfrac{n \sin s}{\sin a \sin b \sin c}.$ (34)

7. ,, $\quad \tan \frac{1}{2} A \tan \frac{1}{2} B \tan \frac{1}{2} C = \dfrac{n}{\sin^2 s}.$ (35)

NOTATIONS.—The arcs which join the vertices of a triangle to the middle points of the opposite sides are called medians, and are denoted by m_a, m_b, m_c, respectively. The arcs of great circles, which are drawn from the vertices at right angles to the opposite sides, are called the altitudes, and represented by h_a, h_b, h_c. The arcs which bisect the interior angles, called the interior bisectors, are denoted by d_a, d_b, d_c; and the bisectors of the exterior angles by d_a', d_b', d_c'.

8. Prove $\quad \cos b + \cos c = 2 \cos \dfrac{a}{2} \cos m_a.$ (36)

9. Prove in a spherical parallelogram that the sum of the cosines of the sides is equal to four times the product of the cosines of the halves of the diagonals.

10. Prove that the norm of the sides of a triangle is equal to the norm of the sides of any of its colunar triangles.

11. If $ABCD$ be a spherical quadrilateral; and if

$$AB = a, \quad CD = a'; \quad BC = b, \quad DA = b'; \quad AC = c, \quad BD = c',$$

and the arcs joining the middle points of a, a'; of b, b'; and c, $c' = \alpha$, β, γ, respectively, it is required to prove

$$\cos a + \cos a' + \cos b + \cos b' = 4 \cos \tfrac{1}{2} c \cos \tfrac{1}{2} c' \cos \gamma. \tag{37}$$

$$\cos b + \cos b' + \cos c + \cos c' = 4 \cos \tfrac{1}{2} a \cos \tfrac{1}{2} a' \cos \alpha. \tag{38}$$

$$\cos c + \cos c' + \cos a + \cos a' = 4 \cos \tfrac{1}{2} b \cos \tfrac{1}{2} b' \cos \beta. \tag{39}$$

12. Prove
$$\cos a + \cos a' + \cos b + \cos b' + \cos c + \cos c' = 2\{\cos \tfrac{1}{2} a \cos \tfrac{1}{2} a' \cos \alpha$$
$$+ \cos \tfrac{1}{2} b \cos \tfrac{1}{2} b' \cos \beta + \cos \tfrac{1}{2} c \cos \tfrac{1}{2} c' \cos \gamma\}. \tag{40}$$

13. Prove
$$\cos a + \cos a' + 4 \cos \tfrac{1}{2} a \cos \tfrac{1}{2} a' \cos \alpha = \cos b + \cos b' + 4 \cos \tfrac{1}{2} b \cos \tfrac{1}{2} b' \cos \beta$$
$$= \cos c + \cos c' + 4 \cos \tfrac{1}{2} c \cos \tfrac{1}{2} c' \cos \gamma. \tag{41}$$

14. Prove

$$\cos^2 \tfrac{1}{2}a + \cos^2 \tfrac{1}{2}a' + 2\cos\tfrac{1}{2}a \cos\tfrac{1}{2}a' \cos\alpha = \cos^2\tfrac{1}{2}b + \cos^2\tfrac{1}{2}b' + 2\cos\tfrac{1}{2}b \cos\tfrac{1}{2}b' \cos\beta$$
$$= \cos^2\tfrac{1}{2}c + \cos^2\tfrac{1}{2}c' + 2\cos\tfrac{1}{2}c \cos\tfrac{1}{2}c' \cos\gamma.$$

15. Given the base of a spherical triangle and the sum of the cosines of the sides, find the locus of the vertex.

16. In a spherical quadrilateral the arcs joining the middle points of opposite sides, and the arc joining the middle points of the diagonals, are concurrent.   (NEUBERG.)

17. If D be any point in the side BC of a spherical triangle, prove that

$$\cos AD \sin BC = \cos AB \sin DC + \cos AC \sin BD. \qquad (42)$$

The theorem of this exercise may be called STEWART'S Theorem.   It is a generalization of a theorem due to that Geometer.—*Sequel to Euclid,* Prop. IX., p. 24.

18. If ABC be an arc of a great circle, and AA', BB', CC', arcs perpendicular to any other great circle, prove that

$$\sin B\bar{C} \sin AA' + \sin CA \sin BB' + \sin AB \sin CC' = 0. \qquad (43)$$

19. Prove $n = \tfrac{1}{2}\sqrt{(1 - \cos^2 a - \cos^2 b - \cos^2 c + 2\cos a \cos b \cos c)}.$

$$(44)$$

20. If c be the diametral side of a diametral triangle, prove

$$\sin^2 \frac{c}{2} = \sin^2 \frac{a}{2} + \sin^2 \frac{b}{2}. \qquad (45)$$

Case II.—Two Sides and the Angles opposite to them.

29. *The sines of the sides of a spherical triangle are proportional to the sines of their opposite angles.*

DEM.—From equations (16), (19) we get, by multiplication,

$$2\sin\tfrac{1}{2}A \cos\tfrac{1}{2}A = \frac{2\sqrt{\sin s \sin(s-a)\sin(s-b)\sin(s-c)}}{\sin b \sin c},$$

or

$$\sin A = \frac{2n}{\sin b \sin c}; \qquad (46)$$

$$\therefore \quad \frac{\sin A}{\sin a} = \frac{2n}{\sin a \sin b \sin c}.$$

In like manner, $\dfrac{\sin B}{\sin b} = \dfrac{2n}{\sin a \sin b \sin c}.$

Hence $\dfrac{\sin A}{\sin a} = \dfrac{\sin B}{\sin b} = \dfrac{\sin C}{\sin c},$ (47)

and the proposition is proved.

Or thus:—Let ABC be the triangle, O the centre of the sphere. From any point D in OA draw DG perpendicular to the plane BOC; and from G draw GE, GF at right angles to OB, OC. Join DE, DF. Now since DG is perpendicular to the plane, and GE perpendicular to OB, a line through G parallel to OB would be perpendicular both to DG and GE,

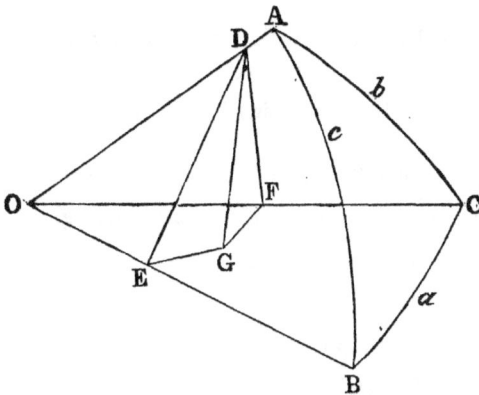

Fig. 13.

and therefore normal to the plane DGE. Hence the angle DEG is equal to the spherical angle B (§ 8). In like manner, DFG is equal to C. Now $DE \sin DEG = DG = DF \sin DFG$; therefore $DE \sin B = DF \sin C$; but $DE = OD \sin DOE = OD \sin c$, and $DF = OD \sin DOF = OD \sin b$. Hence

$$\sin B \sin c = \sin C \sin b,$$

or

$$\sin B : \sin C : : \sin b : \sin c.$$

Exercises.—IV.

1. If a, b, c be the sides of a spherical triangle, a', b', c' the sides of its supplemental triangle, prove

$$\sin a : \sin b : \sin c : : \sin a' : \sin b' : \sin c'. \tag{48}$$

2. Prove that $\quad \sin A \sin B \sin C = \dfrac{8n^3}{\sin^2 a \sin^2 b \sin^2 c}. \tag{49}$

3. Prove that

$$\tan \tfrac{1}{2}(A + B) : \tan \tfrac{1}{2}(A - B) : : \tan \tfrac{1}{2}(a + b) : \tan \tfrac{1}{2}(a - b). \tag{50}$$

4. Prove that

$$\tan \tfrac{1}{2}(A + a) : \tan \tfrac{1}{2}(A - a) : : \tan \tfrac{1}{2}(B + b) : \tan \tfrac{1}{2}(B - b). \tag{51}$$

5. If the bisector AD of the angle A of a spherical triangle divide the side BC into the segments $CD = b'$, $BD = c'$, prove

$$\sin b : \sin c : : \sin b' : \sin c'. \tag{52}$$

6. If the bisector of the exterior angle, formed by producing BA through A, meet the base BC in D', and if $BD' = c''$, $CD' = b''$, prove that

$$\sin b : \sin c : : \sin b'' : \sin c''. \tag{53}$$

7. Prove that

$$\cot \tfrac{1}{2}A : \cot \tfrac{1}{2}B : \cot \tfrac{1}{2}C : : \sin (s - a) : \sin (s - b) : \sin (s - c). \tag{54}$$

8. If D be any point in the side BC of a spherical triangle,

$$\frac{\sin BD}{\sin CD} = \frac{\sin BAD}{\sin CAD} \cdot \frac{\sin C}{\sin B}. \tag{55}$$

Cor.—If D be the middle point,

$$\frac{\sin BAD}{\sin CAD} = \frac{\sin B}{\sin C} = \frac{\sin b}{\sin c}. \tag{56}$$

9. Prove that $\quad \sin a \sin h_a = \sin b \sin h_b = \sin c \sin h_c = 2n. \tag{57}$

10. Given the base of a spherical triangle and the norm of the sides, prove that the locus of the vertex is a small circle.

11. If $m_b = m_c$, prove that either $b = c$, or

$$\sin^2 \tfrac{1}{2} a = \cos^2 \tfrac{1}{2} b + \cos^2 \tfrac{1}{2} c + \cos \tfrac{1}{2} b \cos \tfrac{1}{2} c.$$

12. If a great circle touch two small circles, whose spherical radii are ρ, ρ', and distance between their poles $= \delta$, and if τ denote the arc between the points of contact, prove

$$\sin^2 \tfrac{1}{2} \tau = \frac{\sin^2 \tfrac{1}{2} \delta - \sin^2 \tfrac{1}{2} (\rho - \rho')}{\cos \rho \cos \rho'}. \tag{58}$$

13. If AD be the median, prove

$$\cot ADB = \frac{\cos \frac{1}{2} a \,(\cos b - \cos c)}{2n}. \tag{59}$$

14. Prove that

$$\cot (\overset{\wedge}{m_a,\, a}) \sec \tfrac{1}{2}\, a + \cot (\overset{\wedge}{m_b,\, b}) \sec \tfrac{1}{2}\, b + \cot (\overset{\wedge}{m_c,\, c}) \sec \tfrac{1}{2}\, c = 0. \tag{60}$$

Case III.—Two Sides and two Angles, one of which is contained by the Sides.

30. If we multiply equation (15) by $\cos b$, and substitute the result in (13), we get

$$\cos a = \cos a \cos^2 b + \sin a \sin b \cos b \cos C + \sin b \sin c \cos A.$$

Hence, transposing $\cos a \cos^2 b$, substituting $\sin^2 b$ for $1 - \cos^2 b$, and dividing by $\sin a \sin b$, we get

$$\cot a \sin b = \cos b \cos C + \frac{\sin c}{\sin a} \cos A\,;$$

and substituting, $\quad \dfrac{\sin C}{\sin A}$ for $\dfrac{\sin c}{\sin a}$,

we get $\qquad \cot a \sin b = \cot A \sin C + \cos C \cos b. \tag{61}$

This important formula may be enunciated as follows, calling a the first side, and b the second:—*The cot of the first, by the sine of the second, is equal to the cot of the angle opposite to the first, by the sine of the contained angle, plus the cos of the contained angle, by the cos of the second side.*

By interchanging letters in (61), we get

$$\cot a \sin c = \cot A \sin B + \cos B \cos c. \tag{62}$$

$$\cot b \sin a = \cot B \sin C + \cos C \cos a. \tag{63}$$

$$\cot b \sin c = \cot B \sin A + \cos A \cos c. \tag{64}$$

$$\cot c \sin a = \cot C \sin B + \cos B \cos a. \tag{65}$$

$$\cot c \sin b = \cot C \sin A + \cos A \cos b. \tag{66}$$

Exercises.—V.

1. If D be any point of the side BC, prove that

$$\cot AB \sin DAC + \cot AC \sin DAB = \cot AD \sin BAC. \qquad (67)$$

2. $\qquad \cot ABC \sin DC + \cot ACB . \sin BD = \cot ADB . \sin BC. \qquad (68)$

To prove 1, we have (Art. 29),

$$\cot c \sin AD = \cot D \sin \beta + \cos \beta . \cos AD,$$

$$\cot b \sin AD = - \cot D \sin \gamma + \cos \gamma \cos AD. \qquad \text{Eliminate } \cot D \ldots$$

To prove 2, we have

$$\cot AD \sin m = \cot B \sin D + \cos D \cos m,$$

$$\cot AD \sin n = \cot C \sin D - \cos D \cos n. \qquad \text{Eliminate } \cot AD \ldots$$

3. If we describe a great circle $B'D'C'$, with A as polar, equation (67) gives us

$$\tan DD' . \sin B'C' = \tan BB' . \sin D'C' + \tan CC' . \sin B'D'. \qquad (69)$$

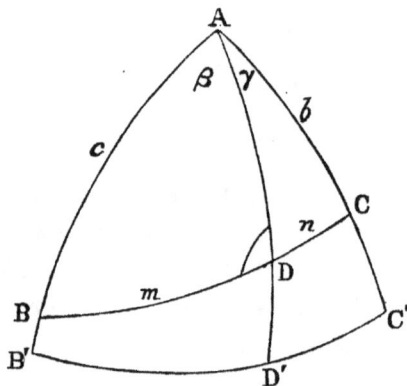

Fig. 14.

4. Prove $\quad 2 \cot m_a . \sin \dfrac{a}{2} = (\cot B + \cot C) \sin (\widehat{m_a . a}). \qquad (70)$

5. ,, $\quad 2 \cos \dfrac{a}{2} \cot (\widehat{m_a . a}) = \cot B - \cot C. \qquad (71)$

6. ,, $\quad \cot m_a . \tan \dfrac{a}{2} = \dfrac{\sin (B + C)}{\sin (B - C)} \cos (\widehat{m_a . a}). \qquad (72)$

7. Prove $\cot b + \cot c = 2 \cos \dfrac{A}{2} \cot d_a.$ (73)

8. ,, $\cos \dfrac{A}{2} \cot d_a + \cos \dfrac{B}{2} \cot d_b + \cos \dfrac{C}{2} \cot d_c = \cot a + \cot b + \cot c.$ (74)

9. ,, $\sin \dfrac{A}{2} \cot d_a' + \sin \dfrac{B}{2} \cot d_b' + \sin \dfrac{C}{2} \cot d_c' = 0.$ (75)

10. ,, $\cos^2 a - \cos^2 b = 2n (\cos a \cot A - \cos b \cot B).$ (76)

Case IV.—Three Angles and a Side.

31. Multiply (61) by $\sin a$, and (63) by $\sin b \cos c$; then

$$\cos a \sin b = \sin a \cot A \sin C + \sin a \cos b \cos C,$$

and

$$\sin a \cos b \cos C = \sin b \cot B \sin C \cos C + \cos a \sin b \cos^2 C.$$

Hence, by substitution and reduction, we get

$$\cos a \sin b \sin C = \sin a \cot A + \sin b \cos B \cos C.$$

Substitute in this expression for $\sin a \cot A$ its equal

$$\frac{\sin b \cos A}{\sin B};$$

reduce, and we get

$$\cos A = - \cos B \cos C + \sin B \sin C \cos a. \qquad (77)$$

Or thus :—Let $A'B'C'$ be the triangle supplemental to ABC; then we have, equation (13),

$$\cos a' = \cos b' \cos c' + \sin b' \sin c' \cos A';$$

but $\qquad a' = \pi - A, \quad b' = \pi - B, \text{ &c. (Art. 15)}.$

Hence, substituting, we get

$$\cos A = - \cos B \cos C + \sin B \sin C \cos a.$$

Interchanging letters in (77), we get

$$\cos B = - \cos C \cos A + \sin C \sin A \cos b. \qquad (78)$$

$$\cos C = - \cos A \cos B + \sin A \sin B \cos c. \qquad (79)$$

32. If we add (77) to the identity

$$\cos (B + C) = \cos B \cos C - \sin B \sin C,$$

we get

$$\sin B \sin C (1 - \cos a) = - \{\cos A + \cos (B + C)\}.$$

Hence (Def. xv.), $2E$ denoting the spherical excess; that is, $A + B + C - \pi$, we get

$$\sin \tfrac{1}{2} a = \sqrt{\frac{\sin E \sin (A - E)}{\sin B \sin C}}. \qquad (80)$$

In like manner,

$$\sin \tfrac{1}{2} b = \sqrt{\frac{\sin E \sin (B - E)}{\sin C \sin A}}, \qquad (81)$$

and

$$\sin \tfrac{1}{2} c = \sqrt{\frac{\sin E \sin (C - E)}{\sin A \sin B}}. \qquad (82)$$

33. If we add (77) to the identity

$$\cos (B - C) = \cos B \cos C + \sin B \sin C),$$

we get $\sin B \sin C (1 + \cos a) = \cos A + \cos (B - C).$

Hence, $\cos \tfrac{1}{2} a = \sqrt{\dfrac{\sin (B - E) \sin (C - E)}{\sin B \sin C}}. \qquad (83)$

In like manner,

$$\cos \tfrac{1}{2} b = \sqrt{\frac{\sin (C - E) \sin (A - E)}{\sin C \sin A}}, \qquad (84)$$

and

$$\cos \tfrac{1}{2} c = \sqrt{\frac{\sin (A - E) \sin (B - E)}{\sin A \sin B}}. \qquad (85)$$

34. From (80) and (83) we get

$$\tan \tfrac{1}{2} a = \sqrt{\frac{\sin E \sin (A - E)}{\sin (B - E) \sin (C - E)}}. \qquad (86)$$

From (81) and (84) we get

$$\tan \tfrac{1}{2} b = \sqrt{\frac{\sin E \sin (B - E)}{\sin (C - E) \sin (A - E)}}. \qquad (87)$$

From (82) and (85) we get

$$\tan \tfrac{1}{2} c = \sqrt{\frac{\sin E \sin (C-E)}{\sin (A-E) \sin B - E)}}. \qquad (88)$$

Cor. 1.—$\tan \tfrac{1}{2} a . \tan \tfrac{1}{2} b = \sin E \div \sin (C-E).$ (89)

$\tan \tfrac{1}{2} b . \tan \tfrac{1}{2} c = \sin E \div \sin (A-E).$ (90)

$\tan \tfrac{1}{2} c . \tan \tfrac{1}{2} a = \sin E \div \sin (B-E).$ (91)

Cor. 2.—$\tan \tfrac{1}{2} a . \cot \tfrac{1}{2} b = \sin (A-E) \div \sin (B-E).$ (92)

$\tan \tfrac{1}{2} b . \cot \tfrac{1}{2} c = \sin (B-E) \div \sin (C-E).$ (93)

$\tan \tfrac{1}{2} c . \cot \tfrac{1}{2} a = \sin (C-E) \div \sin (A-E).$ (94)

Cor. 3.—

$$\tan \tfrac{1}{2} a : \tan \tfrac{1}{2} b : \tan \tfrac{1}{2} c : : \sin (A-E) : \sin (B-E) : \sin (C-E). \qquad (95)$$

EXERCISES.—VI.

1. Prove $\cos C = - \cos (A+B) \cos^2 \tfrac{1}{2} c - \cos (A-B) \sin^2 \tfrac{1}{2} c.$ (96)

2.   ,, $\sin \tfrac{1}{2} a \sin \tfrac{1}{2} b \sin \tfrac{1}{2} c = \dfrac{N \sin E}{\sin A \sin B \sin C},$ (97)

where $N = \sqrt{\sin E \sin (A-E) \sin (B-E) \sin (C-E)}.$ (98)

N is called the *Norm* of the angles of the triangle. *See* note, § 28, Ex. 5.

3. Prove that $\cos \tfrac{1}{2} a \cos \tfrac{1}{2} b \cos \tfrac{1}{2} c = \dfrac{N^2}{\sin E \sin A \sin B \sin C}.$ (99)

4.   ,, $\tan \tfrac{1}{2} a \tan \tfrac{1}{2} b \tan \tfrac{1}{2} c = \dfrac{\sin^2 E}{N}.$ (100)

5.   ,, $\sin a = \dfrac{2N}{\sin B \sin C},$ $\sin b = \dfrac{2N}{\sin C \sin A},$ $\sin C = \dfrac{2N}{\sin A \sin B}.$ (101)

The value $N = \tfrac{1}{2} \sin B \sin C \sin a$, and the corresponding value of n, viz. $\tfrac{1}{2} \sin b \sin c \sin A$, have a remarkable analogy to the equation $S = \tfrac{1}{2} bc \sin A$ in plane trigonometry for the area of a triangle.

6. Prove $\dfrac{\sin A}{\sin a} = \dfrac{\sin B}{\sin b} = \dfrac{\sin C}{\sin c} = \dfrac{N}{n}.$ (102)

7. Prove $2N = \sin A \sin h_a = \sin B \sin h_b = \sin C \sin h_c.$ (103)

8.   ,,   in a right-angled spherical triangle, having the angle C right,

$$\sin \tfrac{1}{2} c = \sqrt{\frac{\sin 2E}{2 \sin A \sin B}}.$$

9.   ,,   in a diametral triangle, having c as the diametral side,

$$\sin \tfrac{1}{2} c = \sqrt{\frac{-\cos (A + B)}{\sin A \sin B}}, \quad \cos \tfrac{1}{2} c = \sqrt{\cot A \cot B}. \quad (104)$$

$$\sin \tfrac{1}{2} a = \sqrt{-\cot B \cot C}, \quad \cos \tfrac{1}{2} a = \sqrt{-\operatorname{cosec} B \cot C}. \quad (105)$$

10. If AD be the bisector of the angle A, prove that

$$\cos B + \cos C = 2 \sin \frac{A}{2} \sin ADB \cos AD. \quad (106)$$

$$\cos C - \cos B = 2 \cos \frac{A}{2} \cos ADB. \quad (107)$$

11. What are the formulae analogous to (106), (107), for the bisector of the external angle ?

Scholium.—In order to pass from a triangle to its polar, it is useful to remark that we replace

$a, b, c, s, A, B, C, s-a, s-b, s-c, A-E, B-E, C-E, n$

by $\pi - A', \pi - B', \pi - C', \pi - E', \pi - a', \pi - b', \pi - c',$

$A' - E', B' - E', C' - E', s' - a', s' - b', s' - c' N.$

In this manner we could infer the formulae of §§ 32, 33 from those of §§ 25, 26.

SECTION II.—FIRST CLASS (continued).

35. The Right-angled Triangle.

The propositions in §§ 24–34 connecting four of the six elements of a spherical triangle assume, in the case of the right-angled triangle, a simpler form when one of the four elements is the right angle, some of the terms vanishing, viz., those containing the cosine or the cotangent of the right angle. These modified formulae are obtained as follows, making the angle C right in each :—

From the equation

$$\frac{\sin A}{\sin a} = \frac{\sin C}{\sin c} \quad (\S\ 29), \text{ we get } \sin A = \frac{\sin a}{\sin c}. \quad (108)$$

From the equation

$$\cot c \sin b = \cot C \sin A + \cos A \cos B \quad (\S\ 30),$$

we get $$\cos A = \frac{\tan b}{\tan c}. \quad (109)$$

From the equation

$$\cot a \sin b = \cot A \sin C + \cos C \cos b \quad (\S\ 30),$$

we get $$\tan A = \frac{\tan a}{\sin b}. \quad (110)$$

From the equation

$$\cos B = -\cos C \cos A + \sin C \sin A \cos b \quad (\S\ 31),$$

we get $$\sin A = \frac{\cos B}{\cos b}. \quad (111)$$

From the equation

$$\cos C = -\cos A \cos B + \sin A \sin B \cos c \quad (\S\ 31),$$

we get $$\cos c = \cot A \cot B. \quad (112)$$

From the equation

$$\cos c = \cos a \cos b + \sin a \sin b \cos C \quad (\S\ 24),$$

we get $$\cos c = \cos a \cos b. \quad (113)$$

36. The formulae (108)–(113) may be proved geometrically as follows :—Let ABC be the triangle, C the right angle, O the centre of the sphere. From any point D in OA erect DF at right angles to OA, meeting OC in F, and draw FE at right angles to OC. Join DE. Then FE is perpendicular to FD, because the plane BOC is perpendicular to AOC. Hence

$$DE^2 = DF^2 + FE^2 = OF^2 - OD^2 + OE^2 - OF^2 = OE^2 - OD^2;$$

therefore the angle ODE is right.

D

Now $\dfrac{FE}{OE} = \dfrac{FE}{ED} \cdot \dfrac{ED}{OE}$; that is, $\sin a = \sin A \sin c.$ (108′)

,, $\dfrac{FD}{OD} = \dfrac{FD}{ED} \cdot \dfrac{ED}{OD}$; ,, $\tan b = \cos A . \tan c.$ (109′)

,, $\dfrac{EF}{OF} = \dfrac{EF}{FD} \cdot \dfrac{FD}{OF}$; ,, $\tan a = \tan A . \sin b.$ (110′)

,, $\dfrac{OD}{OE} = \dfrac{OD}{OF} \cdot \dfrac{OF}{OE}$; ,, $\cos c = \cos a \cos b.$ (113′)

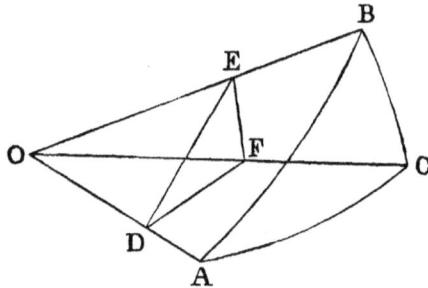

Fig. 15.

Multiply together (110′) and the formula obtained by inter-changing letters, and we get

$$\frac{1}{\cos a \cos b} = \tan A \tan B.$$

Hence $\qquad\qquad \cos c = \cot A \cot B.$ (112′)

Lastly, multiply crosswise (108′) and the formula got from (108′) by interchange of letters; then

$$\sin a \cos B \tan c = \tan a \sin A \sin c ;$$

therefore $\qquad \cos B = \dfrac{\sin A \cos c}{\cos a} = \sin A \cos b.$ (111′)

37. The equations (108)–(111) are easily remembered by comparing them with the corresponding equations for the right-angled plane triangle. Thus—

Plane Triangle.	Spherical Triangle.

$$\sin A = \frac{a}{c},$$

$$\sin A = \frac{\sin a}{\sin c},$$

$$\cos A = \frac{b}{c},$$

$$\cos A = \frac{\tan b}{\tan c},$$

$$\tan A = \frac{a}{b},$$

$$\tan A = \frac{\tan a}{\sin b},$$

$$\sin A = \cos B.$$

$$\sin A = \frac{\cos B}{\cos b}.$$

If in these formulae we interchange the letters A, B, and at the same time the small letters a, b, we get four others, which, however, may be regarded as not essentially different. The formula (113) has a remarkable analogy to Euc. I. xlvii. The formula (112) has no analogue in *Plane Trigonometry*.

38. Napier's Mnemonic Rules. — If as in the annexed diagram we trace in a plane a pentagon, whose sides have respectively the same numerical measures as the quantities

$$a, \; b, \; \frac{\pi}{2} - A, \; \frac{\pi}{2} - c, \; \frac{\pi}{2} - B,$$

we obtain a closed figure representing the system of five quanti-

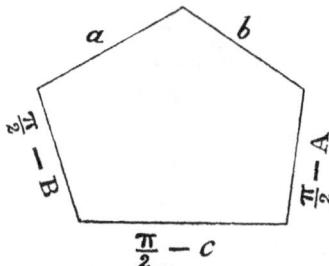

Fig. 16.

ties called *Napier's Circular Parts*. Any one of the five may be selected, and called the *middle part*, then the two next to it are called the *adjacent parts*, and the remaining parts *the opposites*. Thus, if a be selected as the middle part $\frac{\pi}{2} - B$, and b are the adjacents, and $\frac{\pi}{2} - c$, and $\frac{\pi}{2} - A$ the opposites ; then Napier's rules are *sin middle part* = *product of tangents of adjacents* = *product of cosines of opposites*. These rules will be evident from equation (138)–(113). Though given in most treatises on Spherical Trigonometry, they are disapproved of by some of the ablest writers—Delambre, De Morgan, Serret, Baltzer, and others. We have found by experience that the formulae are easily remembered by the method of § 37, which we recommend to the student.

EXERCISES.—VII.

On the right-angled triangle, Ex. 1–20.

1. Prove that $\sin^2 A + \sin^2 b - \sin^2 c = \sin^2 a \sin^2 b.$ \hfill (114)

2. ,, $\sin^2 a \cos^2 b = \sin(c + b) \sin(c - b).$ \hfill (115)

3. ,, $\tan^2 a : \tan^2 b :: \sin^2 c - \sin^2 b : \sin^2 c - \sin^2 a.$ \hfill (116)

4. ,, $\cos^2 A \cdot \sin^2 c = \sin^2 c - \sin^2 a.$ \hfill (117)

5. ,, $\sin^2 A \cos^2 c = \sin^2 A - \sin^2 a.$ \hfill (118)

6. ,, $\sin^2 A \cos^2 b \sin^2 c = \sin^2 c - \sin^2 b.$ \hfill (119)

7. ,, $\cos^2 a \cos^2 B = \sin^2 A - \sin^2 a.$ \hfill (120)

8. ,, $\cos^2 A + \cos^2 c - \cos^2 A \cos^2 c = \cos^2 a.$ \hfill (121)

9. ,, $\sin^2 A - \cos^2 B = \sin^2 a \sin^2 B.$ \hfill (122)

10. ,, $\sin \frac{1}{2} A = \sqrt{\dfrac{\sin(c - b)}{2 \cos b \sin c}}.$ \hfill (123)

11. ,, $\cos \frac{1}{2} A = \sqrt{\dfrac{\sin(c + b)}{2 \cos b \sin c}}.$ \hfill (124)

12. ,, $\sin(a + b) \tan \frac{1}{2}(A + B) = \sin(a - b) \cot \frac{1}{2}(A - B).$ \hfill (125)

13. ,, $\sin(A + B) = \dfrac{\cos a + \cos b}{1 + \cos a \cos b}.$ \hfill (126)

14. ,, $\sin(A - B) = \dfrac{\cos b - \cos a}{1 - \cos a \cos b}.$ \hfill (127)

15. ,, $\cos(A + B) = -\dfrac{\sin a \sin b}{1 + \cos a \cos b}.$ \hfill (128)

16. ,, $\cos(A - B) = \dfrac{\sin a \sin b}{1 - \cos a \cos b}.$ \hfill (129)

17. Prove that $\sin^2 \dfrac{c}{2} = \sin^2 \dfrac{a}{2} \cos^2 \dfrac{b}{2} + \cos^2 \dfrac{a}{2} \sin^2 \dfrac{b}{2}.$ (130)

18.   ,, $\sin (a - b) = \sin a \tan \dfrac{A}{2} - \sin b \tan \dfrac{B}{2}.$ (131)

19.   ,, $\sin (c - b) = \sin (b + c) \tan^2 \dfrac{A}{2},$ (132)

20.   ,, $\sin (c - a) = \cos a \sin b \tan \frac{1}{2} B.$ (133)

21. Prove that in an equilateral spherical triangle

$$2 \sin \tfrac{1}{2} A = \sec \tfrac{1}{2} a. \qquad (134)$$

22. If the opposite angles A, C of a spherical quadrilateral $ABCD$ be right, and if the sides AD, BC produced meet in E, prove

$$\tan AE \,.\, \tan DE = \tan BE \,.\, \tan CE. \qquad (135)$$

23. If the internal and external bisectors of the angle A of a spherical triangle meet the base in D, D', prove

$$\cot DD' = \frac{\sin^2 b - \sin^2 a}{2 \sin a \sin b \sin c}. \qquad (136)$$

24. Given the base of a spherical triangle, and the sum of the base angles, prove that the external bisector of the vertical angle passes through a **given** point.

25. If CC' be the median from the right angle of a right-angled triangle, prove that

$$\sin^2 a + \sin^2 b = \left(2 \cos \frac{c}{2} \sin CC' \right)^2. \qquad (137)$$

26–34. If p be the perpendicular from the right angle C on the hypotenuse, prove—

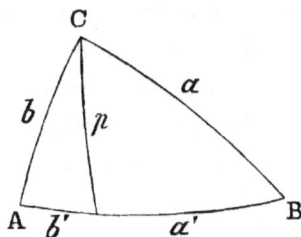

Fig. 17.

1°. $\cot^2 p = \cot^2 a + \cot^2 b.$ 2°. $\cos^2 p = \cos^2 A + \cos^2 B.$ (138)

3°. $\tan^2 a = \mp \tan a' \tan c,$ $\tan^2 b = \pm \tan b' \tan c.$ (139)

4°. $\tan^2 a : \tan^2 b :: \tan a' : \tan b'.$ (140)

5°. $\sin^2 p = \sin a' \sin b'$.   6°. $\sin p \sin c = \sin a \sin b$.   (142)

7°. $\tan a \tan b = \tan c \sin p$.   8°. $\tan^2 a + \tan^2 b = \tan^2 c \cos^2 p$.   (144)

9°. $\cot A : \cot B :: \sin a' : \sin b'$.   (145)

35. If MA', MB', MC' be the perpendiculars let fall from a point M on the sides of the triangle ABC; then

$$\cos AB' . \cos BC' . \cos CA' = \cos A'B . \cos B'C . \cos C'A. \quad \text{(STEINER)}$$
(146)

36. If the triangles ABC, $\alpha\beta\gamma$ be such that perpendiculars let fall from A, B, C on the sides of α, β, γ be concurrent, the perpendiculars from α, β, γ on the sides of ABC are concurrent.

37–40. If AD be the altitude of the triangle ABC, prove—

1°. $\cos BD : \cos CD :: \cos BA : \cos CA$.   (147)

2°. $\sin BD : \sin CD :: \cot B : \cot C$.   (148)

3°. $\tan BD : \tan CD :: \tan BAD : \tan CAD$.   (149)

4°. $\cos BAD : \cos CAD :: \tan BA : \tan CA$.   (150)

41. If the base BC be fixed, and the ratio of the cosines of the sides constant, the locus of A is a great circle perpendicular to BC.

42. If the angle A be fixed, and the ratio $\tan b : \tan c$ constant, the side BC passes through a fixed point.

43. If the base BC be fixed, and the ratio $\tan B : \tan C$ constant, the locus of A is a great circle.

44. If the angle A be fixed, and the ratio $\cos B : \cos C$ constant, the side BC passes through a fixed point.

39. Quadrantal Triangles.

The triangle supplemental to a right-angled triangle has one side a quadrant, and is called a quadrantal triangle. The formulae pertaining to such triangles are obtained from the equations (108)–(113) by the substitutions of the Scholium (Art. 33). They are as follows, c being the quadrantal side:—

$$\sin a = \sin A \div \sin C. \quad (151)$$

$$\cos a = - \tan B \div \tan C. \quad (152)$$

$$\tan a = \tan A \div \sin B. \quad (153)$$

$$\sin a = \cos b \div \cos B. \quad (154)$$

$$\cos C = - \cot a . \cot b. \quad (155)$$

$$\cos C = - \cos A . \cos B. \quad (156)$$

By interchanging the letters a, b, and at the same time the letters A, B in the formulae (151)–(154), we get four others, which, however, may be regarded as not differing essentially from those given.

SECTION III.—SECOND CLASS.

Formulae containing Five Elements.

40. NAPIER'S ANALOGIES.—If we multiply the identity

$$\tan \tfrac{1}{2}(A+B) = \frac{\tan \tfrac{1}{2}A + \tan \tfrac{1}{2}B}{1 - \tan \tfrac{1}{2}A \cdot \tan \tfrac{1}{2}B} \text{ by } \tan \tfrac{1}{2} C,$$

and substitute for $\tan \tfrac{1}{2}A \cdot \tan \tfrac{1}{2} C$, &c., their values, from § 28, *Cor.*, we get

$$\tan \tfrac{1}{2}(A+B) \tan \tfrac{1}{2} C = \frac{\sin(s-b) + \sin(s-c)}{\sin s - \sin(s-a)} = \frac{\cos \tfrac{1}{2}(a-b)}{\cos \tfrac{1}{2}(a+b)}.$$

Hence
$$\tan \tfrac{1}{2}(A+B) = \frac{\cos \tfrac{1}{2}(a-b)}{\cos \tfrac{1}{2}(a+b)} \cot \tfrac{1}{2} C. \tag{157}$$

Similarly,
$$\tan \tfrac{1}{2}(A-B) = \frac{\sin \tfrac{1}{2}(a-b)}{\sin \tfrac{1}{2}(a+b)} \cot \tfrac{1}{2} C. \tag{158}$$

Cor. 1.—If in the expression for $\tan \tfrac{1}{2}(A+B)$ we change *cos* to *sin*, we get the expression for $\tan \tfrac{1}{2}(A-B)$.

41. Again, we have

$$\tan \tfrac{1}{2}(a+b) \cot \tfrac{1}{2} c = \frac{\tan \tfrac{1}{2}a \cot \tfrac{1}{2}c + \tan \tfrac{1}{2}b \cot \tfrac{1}{2}c}{1 - \tan \tfrac{1}{2}a \tan \tfrac{1}{2}b},$$

and substituting for $\tan \tfrac{1}{2}a \cot \tfrac{1}{2}c$, &c., their values from § 34, *Cors.* 1, 2, we get

$$\tan \tfrac{1}{2}(a+b) \cot \tfrac{1}{2}c = \frac{\sin(A-E) + \sin(B-E)}{\sin(C-E) - \sin E} = \frac{\cos \tfrac{1}{2}(A-B)}{\cos \tfrac{1}{2}(A+B)}.$$

Hence
$$\tan \tfrac{1}{2}(a+b) = \frac{\cos \tfrac{1}{2}(A-B)}{\cos \tfrac{1}{2}(A+B)} \tan \tfrac{1}{2} c. \tag{159}$$

Similarly,
$$\tan \tfrac{1}{2}(a-b) = \frac{\sin \tfrac{1}{2}(A-B)}{\sin \tfrac{1}{2}(A+B)} \tan \tfrac{1}{2} c. \tag{160}$$

Cor.—If we change *cos* to *sin* in the expression for tan $\frac{1}{2}(a-b)$ we get that for tan $\frac{1}{2}(a+b)$.

The theorems contained in the equations (157)–(160) may be expressed as proportions, and are called NAPIER'S ANALOGIES, after their discoverer. It may be remarked that the last pair can be got from the first by means of the polar triangle; also that the second and fourth may be inferred from the first and third by multiplying them respectively by the equation

$$\frac{\tan\frac{1}{2}(A-B)}{\tan\frac{1}{2}(A+B)} = \frac{\tan\frac{1}{2}(a-b)}{\tan\frac{1}{2}(a+b)}.$$

Several proofs of these important theorems are known, but the foregoing are probably the simplest.

EXERCISES.—VIII.

1. Show that $\cos a \sin b = \sin a \cos b \cos C + \cos A \sin c$.   (161)

Multiply equation (61) by $\sin a$, and replace $\sin a \cot A \sin C$ by $\cos A \sin c$.

2. Prove that $\sin C \cos a = \cos A \sin B + \sin A \cos B \cos C$.   (162)

SECTION IV.—THIRD CLASS.

Formulae containing Six Elements.

42. DELAMBRE'S ANALOGIES.—1°. To prove

$$\frac{\sin\frac{1}{2}(A+B)}{\cos\frac{1}{2}C} = \frac{\cos\frac{1}{2}(a-b)}{\cos\frac{1}{2}c}.$$

DEM.—$\sin\frac{1}{2}(A+B) = \sin\frac{1}{2}A\cos\frac{1}{2}B + \cos\frac{1}{2}A\sin\frac{1}{2}B$; and substituting for $\sin\frac{1}{2}A$, $\cos\frac{1}{2}A$, &c., their values from §§ 26, 27, we get

$$\sin\frac{1}{2}(A+B) = \frac{\sin(s-b)+\sin(s-a)}{\sin c}\sqrt{\frac{\sin s \cdot \sin(s-c)}{\sin a \cdot \sin b}}.$$

Hence

$$\frac{\sin\frac{1}{2}(A+B)}{\cos\frac{1}{2}C} = \frac{\sin(s-b)+\sin(s-a)}{\sin c} = \frac{\cos\frac{1}{2}(a-b)}{\cos\frac{1}{2}c}. \quad (163)$$

In like manner we get the three following equations :—

2°. $$\frac{\sin \frac{1}{2}(A - B)}{\cos \frac{1}{2}C} = \frac{\sin \frac{1}{2}(a - b)}{\sin \frac{1}{2}c}. \tag{164}$$

3°. $$\frac{\cos \frac{1}{2}(A + B)}{\sin \frac{1}{2}C} = \frac{\cos \frac{1}{2}(a + b)}{\cos \frac{1}{2}c}. \tag{165}$$

4°. $$\frac{\cos \frac{1}{2}(A - B)}{\sin \frac{1}{2}C} = \frac{\sin \frac{1}{2}(a + b)}{\sin \frac{1}{2}c}. \tag{166}$$

From any one of these formulae the others may be obtained by the following rule :—*Change the sign of the letter B (large or small) on one side of the equation, and write* sin *for* cos *and* cos *for* sin *on the other.*

Cor.—Napier's analogies may be inferred from Delambre's by division.

Delambre's analogies were discovered by him in 1807, and published in the *Connaissance des Temps* for 1809, p. 443. They were subsequently discovered independently by GAUSS, and published in his *Theoria motus corporis coelestium in sectionibus conicis solem ambientium.* Both systems may be proved geometrically, though not so directly as by the method in §§ 41, 42. The geometrical proof is the one originally given by Delambre. It was rediscovered by Professor Crofton, F.R.S., in 1869, and published in the *Proceedings of the London Mathematical Society*, Vol. III.

43. REIDT'S ANALOGIES.—From Delambre's analogies we get by an easy transformation four others, due to Reidt. See his *Sammlung von Aufgaben der Trigonometrie*, Seite 233. These may be used in the solution of triangles. Formulae nearly identical with them are given in SERRET'S *Trigonometry*, p. 156. From (163) we get

$$\frac{\cos \frac{1}{2}c - \cos \frac{1}{2}C}{\cos \frac{1}{2}c + \cos \frac{1}{2}C} = \frac{\cos \frac{1}{2}(a - b) - \sin \frac{1}{2}(A + B)}{\cos \frac{1}{2}(a - b) + \sin \frac{1}{2}(A + B)}.$$

Now, put $A + a = 4s',$ $A - a = 4d',$

$B + b = 4s'',$ $B - b = 4d'',$

$C + c = 4s''',$ $C - c = 4d''',$

and we get

$$\tan s''' \tan d''' = \tan \{45° - (s'' + d')\} \tan \{45° - (s' + d'')\}.$$
$$(167)$$

Similarly, from equations (164)–(166), we get

$$\tan (45° + d''') \cot (45° + s''') = \cot (s' - s'') \tan (d' - d'').$$
$$(168)$$

$$\tan (45° + s''') \tan (45° + d''') = \cot (s' + s'') \cot (d' + d'').$$
$$(169)$$

$$\cot s''' \tan d''' = \tan \{45° - (s'' - d')\} \tan \{45° - (s' - d'')\}.$$
$$(170)$$

44. From the formulae (167)–(170) we get, by an obvious method,

$$\tan^2 (45° - s''') = \cot (s' - s'') \tan (s' + s'') \tan (d' - d'') \tan (d' + d'').$$
$$(171)$$

$$\tan^2 (45° - d''') = \tan (s' - s'') \tan (s' + s'') \cot (d' - d'') \tan (d' + d'').$$
$$(172)$$

$$\tan^2 s''' = \tan \{45° - (s'' + d')\} \tan \{45° + (s'' - d')\}$$
$$\tan \{45° - (s' + d'')\} \tan \{45° + (s' - d'')\}. \qquad (173)$$

$$\tan^2 d''' = \tan \{45° - (s'' + d')\} \tan \{45° - (s'' - d')\}$$
$$\tan \{45° - (s' + d'')\} \tan \{45° - (s' - d'')\}. \qquad (174)$$

45. If in the original triangle a, b, B retain their values, and the angle A change into $\pi - A$, the formulae (171), (172) are replaced by the following :—

$$\tan^2 s''' = \tan (s' - s'') \tan (s' + s'') \tan (d' - d'') \cot (d' + d'').$$
$$(175)$$

$$\tan^2 d''' = \tan (s' - s'') \cot (s' + s'') \tan (d' - d'') \tan (d' + d'').$$
$$(176)$$

46. Other Applications of Delambre's Formulae.

From the 3rd and 4th of Delambre's analogies, we have

$$\frac{\cos\left(s-\dfrac{c}{2}\right)}{\cos\dfrac{c}{2}} = \frac{\cos\dfrac{A+B}{2}}{\cos\left(90°-\dfrac{C}{2}\right)} ; \quad \frac{\sin\left(s-\dfrac{c}{2}\right)}{\sin\dfrac{c}{2}} = \frac{\cos\dfrac{A-B}{2}}{\cos\left(90-\dfrac{C}{2}\right)}.$$

From the first of these equations, we get

$$\frac{\cos\left(s-\dfrac{c}{2}\right)+\cos\dfrac{c}{2}}{\cos\dfrac{c}{2}-\cos\left(s-\dfrac{c}{2}\right)} = \frac{\cos\dfrac{A+B}{2}+\cos\left(90-\dfrac{C}{2}\right)}{\cos\left(90-\dfrac{C}{2}\right)-\cos\dfrac{A+B}{2}},$$

or
$$\cot\frac{s}{2}\cot\frac{s-c}{2} = \cot\tfrac{1}{2}E\tan\tfrac{1}{2}(C-E); \qquad (177)$$

and from the second,

$$\frac{\sin\left(s-\dfrac{c}{2}\right)+\sin\dfrac{c}{2}}{\sin\left(s-\dfrac{c}{2}\right)-\sin\dfrac{c}{2}} = \frac{\cos\dfrac{A-B}{2}+\cos\left(90-\dfrac{C}{2}\right)}{\cos\dfrac{A-B}{2}-\cos\left(90-\dfrac{C}{2}\right)},$$

or
$$\tan\frac{s}{2}\cot\frac{s-c}{2} = \cot\tfrac{1}{2}(A-E)\cot\tfrac{1}{2}(B-E). \quad (178)$$

Hence, by division, and extracting the square root, we have

$$\cot\frac{s}{2} = \sqrt{\cot\tfrac{1}{2}E.\tan\tfrac{1}{2}(A-E)\tan\tfrac{1}{2}(B-E)\tan\tfrac{1}{2}(C-E)}.$$
$$(179)$$

By multiplying (177) and (178), we get

$$\tan\frac{s-c}{2} = \sqrt{\tan\tfrac{1}{2}E.\tan\tfrac{1}{2}(A-E)\tan\tfrac{1}{2}(B-E)\cot\tfrac{1}{2}(C-E)}.$$
$$(180)$$

These simple and elegant proofs are due to Prouhet. See *Nouvelles Annales*, 1856, p. 91.

47. If we put

$$L = \sqrt{\tan \tfrac{1}{2} E . \tan \tfrac{1}{2} (A - E) \tan \tfrac{1}{2} (B - E) \tan \tfrac{1}{2} (C - E)},$$

$$(181)$$

equation (179) may be written

$$\cot \frac{s}{2} = \frac{L}{\tan \tfrac{1}{2} E},$$

$$(182)$$

and equation (180) may be written

$$\tan \frac{s - c}{2} = \frac{L}{\tan \tfrac{1}{2} (C - E)}.$$

$$(183)$$

From (183) we get, interchanging letters,

$$\tan \frac{s - a}{2} = \frac{L}{\tan \tfrac{1}{2} (A - E)},$$

$$(184)$$

and

$$\tan \frac{s - b}{2} = \frac{L}{\tan \tfrac{1}{2} (B - E)}.$$

$$(185)$$

48. If we multiply the four equations (182)–(185) we get

$$L = \sqrt{\cot \tfrac{1}{2} s . \tan \tfrac{1}{2} (s - a) \tan \tfrac{1}{2} (s - b) \tan \tfrac{1}{2} (s - c)};$$

$$(186)$$

and substituting this value in (182), we get

$$\tan \tfrac{1}{2} E = \sqrt{\tan \tfrac{1}{2} s . \tan \tfrac{1}{2} (s - a) \tan \tfrac{1}{2} (s - b) \tan \tfrac{1}{2} (s - c)}.$$

$$(187)$$

This beautiful theorem is due to Simon Lhuilier of Geneva. After him I propose to call the function L the *Lhuilierian* of the triangle. It will be seen that on account of its double value, viz., those given in (181) and (186), it will give the solution of a spherical triangle either when the three sides or the three angles are given, that the same system of equations solves both cases, and that in each case the solution is self-verifying.

EXERCISES.—IX.

1. Prove that

$$\sin b \sin c + \cos b \cos c \cos A = \sin B \sin C - \cos B \cos C \cos a. \quad (188)$$

2. Prove Cagnoli's theorem that

$$\cos a \cos B \cos C + \cos A \cos b \cos c = \cos A \cos B \cos C \sin b \sin c$$
$$+ \cos a \cos b \cos c \sin B \sin C. \quad (189)$$

3. Prove $\tan A \tan B \tan C = \dfrac{\tan A}{\cos b \cos c} + \dfrac{\tan B}{\cos b} + \dfrac{\tan C}{\cos c}.$ $\quad (190)$

4. ,, $\tan a \tan b \tan c = \dfrac{\tan a}{\cos B \cos C} - \dfrac{\tan b}{\cos B} - \dfrac{\tan c}{\cos C}.$ $\quad (191)$

5. ,, $\dfrac{\sin A}{\tan b \cos C - \cos A \sin c} = \dfrac{\tan C \cos B + \cos a \sin B}{\sin a}.$ $\quad (192)$

6. ,, $\sin C \cos b = \cos B \sin A + \cos A \sin b \cos C.$ $\quad (193)$

7. ,, $\sin c \cos B = \cos b \sin a - \cos a \sin b \cos C.$ $\quad (194)$

8. ,, $\dfrac{\tan A \sin B - \cos b \cos c}{\sin C} = \dfrac{\tan a \tan b + \cos C}{\tan b - \tan a \cos C}.$ $\quad (195)$

9. ,, $\dfrac{\tan a \sin b + \cos b \cos C}{\sin C} = \dfrac{\tan A \tan B - \cos c}{\tan B + \tan A \cos c}.$ $\quad (196)$

10. ,, $\dfrac{\sin C}{\sin B \tan c} = \sin a \cos C + \cos a \cot b.$ $\quad (197)$

11. ,, $\dfrac{\sin C}{\sin b \tan C} = \sin A \cos c - \cos A \cot B.$ $\quad (198)$

The Exercises 3–11 are due to Barbier. See *Nouvelles Annales* for 1866, p. 349. The following is an outline of the method of proof:—

ABC, $A'B'C'$ are two supplemental triangles.

To prove 3—$A'AI$ is perpendicular to BC.

Hence $\cos c = \cot B \cot \beta, \quad \cos b = \cot C \cot \gamma, \quad A = \beta + \gamma;$

$\therefore \quad \tan A = \tan \beta + \tan \gamma + \tan A \tan \beta \tan \gamma,$ &c.

To prove 1—Compare HK in the triangles AHK, $A'HK$.

,, 5—Apply the formula (61) to the triangles $A'HK$, AHK.

,, 6—Compare DC in the triangles DCE, DBC.

To prove 8—$\tan GAB = \dfrac{\tan GAC - \tan A}{1 + \tan GAC \cdot \tan A}$ . . .

,,   10—The angles AFC, AFK, are complementary, &c.

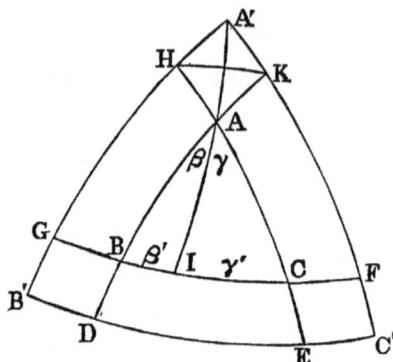

Fig. 18.

The Exercises 4, 7, 9, 11 are inferred from 3, 6, 8, 10 by the properties of supplemental triangles.

12. In a right-angled triangle, prove $\sin c = \dfrac{n}{N}$.   (199)

13. If the sides of a spherical quadrilateral be α, β, γ, δ, and the diagonals Σ and ϕ, prove that

$$\cos (\overset{\wedge}{\Sigma\phi}) = (\cos \alpha \cos \gamma - \cos \beta \cos \delta) \operatorname{cosec} \Sigma \cdot \operatorname{cosec} \phi. \quad (200)$$

14. Prove that

$$\cos (\overset{\wedge}{\alpha\gamma}) = (\cos \beta \cos \delta - \cos \Sigma \cdot \cos \phi) \operatorname{cosec} \alpha \cdot \operatorname{cosec} \gamma. \quad (201)$$

15. Prove that

$$\cos (\overset{\wedge}{\beta\delta}) = (\cos \alpha \cos \gamma - \cos \Sigma \cos \phi) \operatorname{cosec} \beta \operatorname{cosec} \delta. \quad (202)$$

16. Prove that the angle between the bisector of the angle C of a spherical triangle ABC and the perpendicular from C on AB

$$= \tan^{-1}\left\{ \tan \tfrac{1}{2} (A + B) \tan \tfrac{1}{2} (A - B) \tan \frac{c}{2}\right\}. \quad (203)$$

17.   ,, $(\cos^2 A - \cos^2 B) \div (\cot a \cos A - \cot b \cos B) = 2N.$ (204)

18.   ,, $\sin A \sin B \sin C \sin a \sin b \sin c = 4nN.$ (205)

19.   ,, $N = 2n^2 \div \sin a \sin b \sin c.$ (206)

20.   ,, $\dfrac{N^2}{n^2} = \dfrac{1 + \cos A \cos B \cos C}{1 - \cos a \cos b \cos c}.$ (207)

21. Prove that the sines of the perpendiculars from the orthocentre on the sides of a spherical triangle are proportional to the secants of the opposite angles.

22. If the base BC of a spherical triangle be given in position and magnitude, and the sum of the sides AB, AC be given in magnitude, prove that the locus of the intersection of the bisector of the external vertical with a great circle perpendicular to AB at B is a great circle.

23–30. Prove the following analogies due to Breitschneider. See CRELLE's *Journal*, Band. xiii., Seite 145 :—

$$1°. \quad \frac{\sin \frac{1}{2} E . \cos \frac{1}{2} (A - E)}{\sin \frac{1}{2} A} = \frac{\sin \frac{1}{2} s . \sin \frac{1}{2} (s - a)}{\cos \frac{1}{2} a}. \tag{208}$$

$$2°. \quad \frac{\cos \frac{1}{2} E . \sin \frac{1}{2} (A - E)}{\sin \frac{1}{2} A} = \frac{\cos \frac{1}{2} s . \cos \frac{1}{2} (s - a)}{\cos \frac{1}{2} a}. \tag{209}$$

$$3°. \quad \frac{\sin \frac{1}{2} E . \sin \frac{1}{2} (A - E)}{\cos \frac{1}{2} A} = \frac{\sin \frac{1}{2} (s - b) \sin \frac{1}{2} (s - c)}{\cos \frac{1}{2} a}. \tag{210}$$

$$4°. \quad \frac{\cos \frac{1}{2} E . \cos \frac{1}{2} (A - E)}{\cos \frac{1}{2} A} = \frac{\cos \frac{1}{2} (s - b \cos \frac{1}{2} (s - c)}{\cos \frac{1}{2} a}. \tag{211}$$

$$5°. \quad \frac{\cos \frac{1}{2} (B - E) \cos \frac{1}{2} (C - E)}{\sin \frac{1}{2} A} = \frac{\sin \frac{1}{2} s . \cos \frac{1}{2} (s - a)}{\sin \frac{1}{2} a}. \tag{212}$$

$$6°. \quad \frac{\sin \frac{1}{2} (B - E) \sin \frac{1}{2} (C - E)}{\sin \frac{1}{2} A} = \frac{\cos \frac{1}{2} s . \sin \frac{1}{2} (s - a)}{\sin \frac{1}{2} a}. \tag{213}$$

$$7°. \quad \frac{\cos \frac{1}{2} (B - E) \sin \frac{1}{2} (C - E)}{\cos \frac{1}{2} A} = \frac{\sin \frac{1}{2} (s - b) \cos \frac{1}{2} (s - c)}{\sin \frac{1}{2} a}. \tag{214}$$

$$8°. \quad \frac{\sin \frac{1}{2} (B - E) \cos \frac{1}{2} (C - E)}{\cos \frac{1}{2} A} = \frac{\cos \frac{1}{2} (s - b) \sin \frac{1}{2} (s - c)}{\sin \frac{1}{2} a}. \tag{215}$$

These analogies are all inferences from Delambre's. For example, 1° is obtained by subtracting both sides of the equation

$$\frac{\cos \frac{1}{2} (B + C)}{\sin \frac{1}{2} A} = \frac{\cos \frac{1}{2} (b + c)}{\cos \frac{1}{2} a} \quad \text{from unity.}$$

CHAPTER III.

SOLUTION OF SPHERICAL TRIANGLES.

49. Preliminary Observations.

1°. The logarithms of trigonometrical functions are obtained from their "Tabular Logarithms" by subtracting 10 from the characteristics. For example,

$$\log \tan 37° \ 40' \ 16'' = \bar{1} . 8876649.$$

The ablest recent continental writers, such as SERRET, BRIOT, et BOUQUET, and others, employ the logarithms thus reduced, instead of the Tabular Logarithms. We may add that the late PROF. BOOLE was strongly in favour of this alteration.

2°. It is necessary to avoid the calculation of very small angles by their cosines, or of angles near 90° by their sines, for their tabular differences vary too slowly. It is better to determine such angles, for example, by means of their tangents.

3°. When angles greater than 90° occur in calculation, we replace them by their supplements, and if the functions of such angles be either *cos*, *sec*, *tan* or *cot*, we take account of the change of sign.

4°. Formulae not adapted to logarithmic computation can be rendered so by means of an auxiliary angle. Thus :—

(*a*) For $A \cos a + B \sin a$ we put $A = B \tan \phi$,

which gives
$$\frac{B \sin (\phi + a)}{\cos \phi}. \qquad (216)$$

(*b*) For $A \cos a + B$ we put $B = A \sin a \tan \phi$,

which gives
$$\frac{A \cos (a - \phi)}{\cos \phi}. \qquad (217)$$

Section I.—The Right-angled Triangle.

50. The solution of right-angled spherical triangles presents six distinct cases, which correspond to and are solved by the six equations (108)–(113) of § 35. For their discussions the following remarks are useful :—1°. *The three sides of a spherical triangle* (omitting triangles birectangular or trirectangular), *are either all acute, or else one is acute and the other two obtuse.* This follows from the equation $\cos c = \cos a \cos b$. 2°. *Either of the sides containing the right angle is of the same species as the opposite angle.* This can be inferred from the equation

$$\cos A = \cos a \sin B.$$

It will be a useful exercise for the student to prove these propositions geometrically.

51. First Case.—*Being given c and a, to calculate b, A, B.*

The required parts are given by the formulae

$$\cos b = \frac{\cos c}{\cos a}, \text{ equation (113).} \qquad (218)$$

$$\sin A = \frac{\sin a}{\sin c}, \quad \text{,,} \quad (108). \qquad (219)$$

$$\cos B = \frac{\tan a}{\tan c}, \quad \text{,,} \quad (109). \qquad (220)$$

The formula (219) gives two supplementary values of $\sin A$, but the ambiguity is removed by considering that A must be of the same species as a.

From the equations (218)–(220) we get, by obvious transformations, the three following :—

$$\tan \tfrac{1}{2} b = + \sqrt{\tan \tfrac{1}{2}(c+a)\tan \tfrac{1}{2}(c-a)}. \qquad (221)$$

$$\tan(45 + \tfrac{1}{2}A) = \pm \sqrt{\frac{\tan \tfrac{1}{2}(c+a)}{\tan \tfrac{1}{2}(c-a)}}. \qquad (222)$$

$$\tan \tfrac{1}{2} B = + \sqrt{\frac{\sin(c-a)}{\sin(c+a)}}. \qquad (223)$$

E

The equation (221) proves that if $\frac{1}{2}(c+a)$ be greater than 90°, c is less than a, for the product of the quantities under the radical must be +. The sign is + or − in (222), according as a is less or greater than 90°. If the given parts c and a be each 90°, the angle A is 90°, and b is indeterminate. It is also evident, from the formulae (218)–(220), that c must lie between a and $\pi - a$, in order that the values of $\cos b$, $\cos B$, and $\sin A$ may be numerically less than unity.

EXAMPLE—

Given $c = 37° \ 40' \ 20''$, $a = 37° \ 40' \ 12''$; find b, A, B.

Type of the Calculation.

$c + a = 75° \ 20' \ 32''$, $\frac{1}{2}(c+a) = 37° \ 40' \ 16''$.

$c - a = \ 0 \quad 0 \quad 8$, $\frac{1}{2}(c-a) = \ 0 \quad 0 \quad 4$.

$l \tan \frac{1}{2}(c+a) = \overline{1}\cdot8876649$, $l \sin (c+a) = \overline{1}\cdot9856305$,

$l \tan \frac{1}{2}(c-a) = \overline{5}\cdot2876348$; $l \sin (c-a) = \overline{5}\cdot5886648$;

$\therefore \ l \tan \frac{1}{2} b = \overline{3}\cdot5876498$. $\therefore \ l \tan (45 + \frac{1}{2} A) = 2\cdot3000150$,

$l \tan \frac{1}{2} B = \overline{3}\cdot8015172$.

Hence $b = 0° \ 26' \ 37''\cdot2$, $A = 89° \ 25' \ 37''$, $B = 0° \ 43' \ 33''$.

EXERCISES.—X.

1. Given $c = 63° \ 55' \ 43''$, $a = 120° \ 10' \ 0''$; find b, A, B.
2. ,, $c = 54 \quad 20 \quad 0$, $a = \ 36 \quad 27 \quad 0$; ,, ,,
3. ,, $c = 87 \quad 11 \quad 39\cdot8$, $a = \ 86 \quad 40 \quad 0$; ,, ,,

52. SECOND CASE.—*Being given c, A, to calculate a, b, B.*

The unknown parts are found thus :—

$$\sin a = \sin c \ \sin A, \quad \text{equation (108).} \qquad (224)$$

$$\tan b = \tan c \ \cos A, \qquad ,, \quad (109). \qquad (225)$$

$$\cot B = \cos c \ \tan A, \qquad ,, \quad (112). \qquad (226)$$

The ambiguity in finding a by its sine is removed in § 51. If a be very near 90°, we commence by calculating the values of b and B, and then determine a by either of the formulae

$$\tan a = \sin b \tan A. \qquad (227)$$

$$\tan a = \tan c \cos B. \qquad (228)$$

EXAMPLE—

Given $c = 81° 29' 32''$, $A = 32° 28' 17''$; find a, b, B.

Type of the Calculation.

$c = 81° 29' 32''$,	$A = 32° 28' 17''$.
$l \sin c = \overline{1}\cdot 9951945$	$l \tan c = \cdot 8250982$
$l \sin A = 1\cdot 7278843$	$l \cos A = 1\cdot 9269687$

$l \sin a = \overline{1}\cdot 7230788$;	$l \tan b = \cdot 7520669$;
$\therefore\ a = 31° 54' 25''.$	$\therefore\ b = 79° 51' 48''\cdot 65.$

$$l \cos c = \overline{1}\cdot 1700960$$

$$l \tan A = \overline{1}.8009157$$

$$l \cot B = \overline{2}\cdot 9710117;$$

$$\therefore\ B = 84° 39' 21''\cdot 33.$$

EXERCISES.—XI.

1. Given $c = 69° 25' 11''$, $A = 54° 54' 42''$; find a, b, B.
2. ,, $c = 112\ \ 48\ \ 0$, $A = 56\ \ 11\ \ 56$; ,, ,,
3. ,, $c = 46\ \ 40\ \ 12$, $A = 37\ \ 46\ \ 9$; ,, ,,

53. THIRD CASE.—*Being given a, b, to find A, B, c.*

The formulae are—

$$\tan A = \tan a \div \sin b, \text{ equation (110).} \qquad (229)$$

$$\tan B = \tan b \div \sin a, \quad \text{,,} \quad (110). \qquad (230)$$

$$\cos c = \cos a \cos b, \quad \text{,,} \quad (113). \qquad (231)$$

If the side c be very small, instead of the formulae (231), we may determine it by means of either of the following equations—

$$\tan c = \tan b \div \cos A = \tan a \div \cos B, \qquad (232)$$

A, B having been calculated from the formulae (229), (230).

Exercises.—XII.

1. Given $a = 120°\ 10'\ 0''$, $b = 150°\ 59'\ 44''$;   find A, B, c.

2.   ,, $a = 36\ \ 27\ \ 0$, $b = 43\ \ 32\ \ 31$;   ,,   ,,

3.   ,, $a = 86\ \ 40\ \ 0$, $b = 32\ \ 40\ \ 0$;   ,,   ,,

54. Fourth Case.—*Being given a, A, to find c, b, B.*

In this case we have

$$\sin c = \sin a \div \sin A, \text{ equation (108)}. \qquad (233)$$
$$\sin b = \tan a \div \tan A, \quad \text{,,} \quad (110). \qquad (234)$$
$$\sin B = \cos A \div \cos a, \quad \text{,,} \quad (111). \qquad (235)$$

Since each of the sought parts is found by its sine, there will be on the whole six solutions, the formulae (233)–(235) giving two values for each of the sought parts c, b, B. The parts a, A must be of the same species (see § 50). Also $\sin a$ must be less than $\sin A$ (A is comprised between a and 90), and the formula (233) gives two admissible values of c, say c_1 and $180 - c_1$; the formula (234) gives two values of b, b_1 and $180 - b_1$, one of which goes with c_1, and the other with $180 - c_1$; for the three sides a, b, c are all acute, or one alone is acute; the formula (235) gives two values of B, of which one goes with b_1, and the other with $180 - b_1$ (b and B are of the same species). If the sought parts are badly determined by the equations (233)–(235), which happens when they are near 90°, the preceding formulae may be written as follows (see Notation, § 43):—

$$\tan\left(45° + \tfrac{1}{2}c\right) = \pm\sqrt{\frac{\tan 2s'}{\tan 2d'}}. \tag{236}$$

$$\tan\left(45° + \tfrac{1}{2}b\right) = \pm\sqrt{\frac{\sin 4s'}{\sin 4d'}}. \tag{237}$$

$$\tan\left(45° + \tfrac{1}{2}B\right) = \pm\sqrt{\frac{\cot 2s'}{\cot 2d'}}. \tag{238}$$

Each of the radicals on the right-hand side must have the double sign.

EXERCISES.—XIII.

1. Given $a = 34° \ 6' \ 13''$, $A = 34° \ 7' \ 41''$; find c, b, B.
2. ,, $a = 87 \ 12 \ 28$, $A = 87 \ 51 \ 37$; ,, ,,

55. FIFTH CASE.—*Being given a, B, to find b, A, c.*

The following equations give the required parts—

$$\tan b = \sin a \tan B, \text{ equation (110).} \tag{239}$$
$$\cos A = \cos a \sin B, \quad ,, \quad (111). \tag{240}$$
$$\tan c = \tan a \div \cos B \quad ,, \quad (109). \tag{241}$$

If A be small instead of the equation (240), the following may be used—

$$\tan A = \frac{\tan a}{\sin b}, \text{ equation (110).} \tag{242}$$

EXERCISES.—XIV.

1. Given $a = 92° \ 47' \ 32''$, $B = 50° \ 2' \ 1''$; find b, A, c.
2. ,, $a = 96 \ 49 \ 59$, $B = 50 \ 12 \ 4$; ,, ,,
3. ,, $a = 20 \ 20 \ 20$, $B = 38 \ 10 \ 10$; ,, ,,

56. SIXTH CASE.—*Being given A, B, to find a, b, c.*

Here we have

$$\cos a = \cos A \div \sin B, \text{ equation (111).} \tag{243}$$
$$\cos b = \cos B \div \sin A, \quad ,, \quad (111). \tag{244}$$
$$\cos c = \cot A \cdot \cot B, \quad ,, \quad (112). \tag{245}$$

If these formulae be not well adapted for the arithmetical calculation, which happens when the sought parts are small, we use the following—

$$\tan \tfrac{1}{2} a = + \sqrt{\frac{\tan E}{\tan (B - E)}}. \tag{246}$$

$$\tan \tfrac{1}{2} b = + \sqrt{\frac{\tan E}{\tan (A - E)}}. \tag{247}$$

$$\tan \tfrac{1}{2} c = + \sqrt{\frac{\sin E}{\cos (A - B)}}. \tag{248}$$

Exercises.—XV.

1. Given $A = 63° \ 15' \ 12''$, $B = 163° \ 33' \ 39''$; find $a \ b, \ c$.
2. „ $A = 46 \quad 59 \quad 42$, $B = \ 57 \quad 59 \quad 17$;   „   „
3. „ $A = 42 \ \vdots \ 24 \quad 9$, $B = \ 99 \quad 4 \quad 11$;   „   „

57. The triangle supplemental to a right-angled triangle is a quadrantal spherical triangle, that is, a triangle one of whose sides is a quadrant. The solution can be inferred from the equations of § 39. Other triangles besides the quadrantal can be reduced to the rectangular. 1°. Isosceles triangles, for the median that bisects the base is perpendicular to it. 2°. Triangles in which $a + b = \pi$, or $A + B = \pi$, for the colunar triangle $B'AC$ is isosceles.

Section II.—Oblique-angled Triangles.

58. The solution of oblique-angled triangles presents three pairs of cases, each consisting of two which are reciprocals of each other. They are as follows:—

I.—*The three sides and its reciprocal the three angles.*

II.—*Two sides and the angle opposite to one of them; its reciprocal two angles and the side opposite to one of them.*

III.—*Two sides and their included angle; its reciprocal two angles and the adjacent side.*

First Pair of Cases.

59. *Being given the three sides a, b, c, to calculate the angles.*

From equation (186) and equations (182)–(185), we get

$$\log L = \tfrac{1}{2} \{ l \cot \tfrac{1}{2} s + l \tan \tfrac{1}{2} (s-a) + l \tan \tfrac{1}{2}(s-b) + l \tan \tfrac{1}{2}(s-c) \}.$$
$$(249)$$

$$l \tan \tfrac{1}{2} E = \log L - l \cot \tfrac{1}{2} s. \qquad (250)$$

$$l \tan \tfrac{1}{2}(A - E) = \log L - l \tan \tfrac{1}{2}(s-a). \qquad (251)$$

$$l \tan \tfrac{1}{2}(B - E) = \log L - l \tan \tfrac{1}{2}(s-b). \qquad (252)$$

$$l \tan \tfrac{1}{2}(C - E) = \log L - l \tan \tfrac{1}{2}(s-c). \qquad (253)$$

EXAMPLE—

Given $a = 100°$, $b = 37° 18'$, $c = 62° 46'$; find A, B, C.

Type of the Calculation.

$a = 100° \ 0' \ 0''$	$l \cot \tfrac{1}{2} s \quad = \bar{1}{\cdot}9235570$
$b = 37 \ 18 \ 0$	$l \tan \tfrac{1}{2}(s-a) = \bar{4}{\cdot}4637261$
$c = 62 \ 46 \ 0$	$l \tan \tfrac{1}{2}(s-b) = \bar{1}{\cdot}7150481$
$\therefore \quad \tfrac{1}{2} s = 50 \ 1 \ 0$	$l \tan \tfrac{1}{2}(s-c) = \bar{1}{\cdot}5278682$
$\tfrac{1}{2}(s-a) = 0 \ 1 \ 0$	$2 \,\lfloor\, \bar{5}{\cdot}7001994$
$\tfrac{1}{2}(s-b) = 31 \ 22 \ 0$	$\therefore \quad \log L = \bar{3}{\cdot}8500997$
$\tfrac{1}{2}(s-c) = 18 \ 38 \ 0$	Hence $\quad l \tan \tfrac{1}{2} E = \bar{3}{\cdot}9265427$

$$,, \quad l \tan \tfrac{1}{2}(A-E) = 1{\cdot}3863736$$
$$,, \quad l \tan \tfrac{1}{2}(B-E) = \bar{2}{\cdot}0650516$$
$$,, \quad l \tan \tfrac{1}{2}(C-E) = \bar{2}{\cdot}3222315$$

$$\therefore \ E = 0° 58' 3''{\cdot}32, \quad A = 176° 15' 46''{\cdot}56, \quad B = 2° 17' 55''{\cdot}08,$$
$$C = 3° 22' 25''{\cdot}46.$$

EXERCISES.—XVI.

1. Given $a = 89° \, 59' \, 59'$, $b = 88° \, 55' \, 58''$, $c = 87° \, 57' \, 57''$; find A, B, C.
2.   ,, $a = 120 \; 55 \; 35$, $b = 59 \;\; 4 \; 25$, $c = 106 \; 10 \; 22$;   ,,   ,,
3.   ,, $a = \;\, 20 \; 16 \; 38$, $b = 56 \; 19 \; 40$, $c = \;\, 66 \; 20 \; 44$;   ,,   ,,

60. If only one angle, say A, be required, the formula for $\tan \frac{1}{2} A$ in terms of the sides is simpler than the foregoing. Thus in logarithms,

$$l \tan \tfrac{1}{2} A = \tfrac{1}{2} \{ l \sin (s - b) + l \sin (s - c) - l \sin s - l \sin (s - a) \}. \tag{254}$$

$$
\begin{array}{ll}
a = 82° \; 33' \; 51'' & l \sin (s - b) = \overline{1} \cdot 9788195 \\
b = 27 \quad 16 \quad\;\, 9 & l \sin (s - c) = \overline{1} \cdot 2529286 \\
\underline{c = 89 \quad 12 \quad 24} & \overline{\overline{1} \cdot 2317481} \\
s = 99 \quad 31 \quad 12 & l \sin s \quad\;\; = \overline{1} \cdot 9939773 \\
s - a = 16 \quad 57 \quad 21 & l \sin (s - a) = \overline{1} \cdot 4648388 \\
s - b = 72 \quad 15 \quad\;\, 3 & \overline{\overline{1} \cdot 4588161} \\
s - c = 10 \quad 18 \quad 48 & \big[\overline{\overline{1} \cdot 7729320}
\end{array}
$$

$$\therefore \quad l \tan \tfrac{1}{2} A \;\; = \overline{1} \cdot 8864660$$

$$\text{Hence} \qquad A = 75° \; 11' \; 22''.$$

61. *Being given the three angles A, B, C, to calculate the sides.*

From equations (181)–(185) we have

$$\log L = \tfrac{1}{2} \{ l \tan \tfrac{1}{2} E + l \tan \tfrac{1}{2} (A - E) + l \tan \tfrac{1}{2} (B - E) + l \tan \tfrac{1}{2} (C - E) \}. \tag{255}$$

$$l \cot \tfrac{1}{2} s = \log L - l \tan \tfrac{1}{2} E. \tag{250'}$$

$$l \tan \tfrac{1}{2} (s - a) = \log L - l \tan \tfrac{1}{2} (A - E). \tag{251'}$$

$$l \tan \tfrac{1}{2} (s - b) = \log L - l \tan \tfrac{1}{2} (B - E). \tag{252'}$$

$$l \tan \tfrac{1}{2} (s - c) = \log L - l \tan \tfrac{1}{2} (C - E). \tag{253'}$$

On comparing the equations (250')–(253') with (250)–(253) of § 59, it will be seen that they are identical

1. Given $A = 161°\ 22'\ 10''$, $B = 26°\ 58'\ 46''$, $C = 39°\ 45'\ 10''$;

find a, b, c.

2. ,, $A = 127\ \ 22\ \ \ 7$, $B = 128\ \ 41\ \ 49$, $C = 107\ \ 33\ \ 20$;

find a, b, c.

3. ,, $A = \ 78\ \ 15\ \ 41$, $B = 153\ \ 17\ \ \ 6$, $C = \ 87\ \ 43\ \ 36$;

find a, b, c.

62. If only one side, say a, be required, we may use either of the formulae

$$\tan \tfrac{1}{2} a = \sqrt{\frac{\sin E . \sin (A - E)}{\sin (B - E) \sin (C - E)}},$$

or
$$\cos a = \frac{\cos A + \cos B \cos C}{\sin B \sin C}.$$

The last can be adapted to logarithmic computation by means of an auxiliary angle. Thus, if we put

$$\tan \phi = \frac{\sin B \cos C}{\cos A},$$

we get
$$\cos a = \frac{\sin (B + \phi) \cot C}{\sin B \sin \phi}. \qquad (256)$$

EXAMPLE.—Given

$A = 32°\ 54'\ 28''$, $B = 146°\ 58'\ 9''$, $C = 24°\ 54'\ 47''$; find a.

$l \sin B = \bar{1}\cdot7364682$	$l \cot C \quad\ \ = 0\cdot3330492$
$l \cos C = \bar{1}\cdot9575824$	$l \sin (B + \phi) = \bar{2}\cdot6464053$
$\qquad\quad\ \ \bar{1}\cdot6940506$	$\qquad\qquad\quad\ \ \bar{2}\cdot9794545$
$l \cos A = \bar{1}\cdot9240447$	$l \sin B \quad\ \ = \bar{1}\cdot7364682$
$\therefore\quad l \tan \phi = \bar{1}\cdot7700059$	$l \sin \phi \quad\ \ = \bar{1}\cdot7053630$
Hence $\quad \phi = 30°\ 29'\ 30''\cdot4$	$\qquad\qquad\quad\ \ \bar{1}\cdot4418312$
$\therefore\quad B + \phi = 177\ \ 27\ \ 39\cdot4.$	$\therefore\ l \cos a \quad\ = \bar{1}\cdot5376233.$

Hence $a = 69°\ 49'\ 40''$.

Second Pair of Cases.

63. *Given two sides a, b, and the angle A opposite to one of them, to calculate the remaining parts.*

The sought parts are found by the following equations:—

$$\sin B = \frac{\sin A \cdot \sin b}{\sin a}. \tag{257}$$

$$\tan \tfrac{1}{2} c = \tan \tfrac{1}{2} (a - b) \frac{\sin \tfrac{1}{2}(A + B)}{\sin \tfrac{1}{2}(A - B)}. \tag{258}$$

$$\tan \tfrac{1}{2} C = \cot \tfrac{1}{2}(A - B) \frac{\sin \tfrac{1}{2} (a - b)}{\sin \tfrac{1}{2} (a + b)}. \tag{259}$$

The formula (257) gives for B two values B_1, $180° - B_1$, if $\sin A \sin b$ be less than $\sin a$. In order that either of these may be admissible, it is necessary and sufficient that, when substituted in (258), (259), they give positive values for $\tan \tfrac{1}{2} c$ and $\tan \tfrac{1}{2} C$, or, which is the same thing, that $a - b$ and $A - B$ will be of the same sign. This condition is both *necessary* and *sufficient.* For a, b, A, being the given elements, denote by B, c, C the other elements determined by the equations (257)–(259). Now let us construct a triangle T, having the angle C and the sides a, b, and calling A', B', c' the other elements of this triangle, we have

$$\tan \tfrac{1}{2} c' = \tan \tfrac{1}{2} (a - b) \frac{\sin \tfrac{1}{2} (A' + B')}{\sin \tfrac{1}{2} (A' - B')}. \tag{258'}$$

$$\tan \tfrac{1}{2} C = \cot \tfrac{1}{2}(A' - B') \frac{\sin \tfrac{1}{2} (a - b)}{\sin \tfrac{1}{2} (a + b)}. \tag{259'}$$

$$\frac{\tan \tfrac{1}{2} (A' + B')}{\tan \tfrac{1}{2} (A' - B')} = \frac{\tan \tfrac{1}{2} (a + b)}{\tan \tfrac{1}{2} (a - b)}, \tag{a}$$

and from (257) we infer

$$\frac{\tan \tfrac{1}{2} (A + B)}{\tan \tfrac{1}{2} (A - B)} = \frac{\tan \tfrac{1}{2} (a + b)}{\tan \tfrac{1}{2} (a - b)}. \tag{a}$$

If we compare (259) and (259'), we see that $A' - B' = A - B$. Hence, from the two formulae (α) we have $A' + B' = A + B$; therefore $A' = A$, $B' = B$. Lastly, from (258) and (258') we infer that $c' = c$. Hence we have the following Rule :—*If each of the two values of B which are got from* (257) *be such as that* $(A - B)$ *and* $(a - b)$ *have like signs, there are two solutions. If only one of them satisfies this condition, there is only one triangle that satisfies the problem. The problem is impossible when neither of the values of B make* $(A - B)$ *and* $(a - b)$ *of the same sign.*

Instead of the formulae (258), (259), we may use the following :—

$$l \tan \tfrac{1}{2} c = l \tan \tfrac{1}{2} (a + b) + l \cos \tfrac{1}{2} (A + B) - l \cos \tfrac{1}{2} (A - B).$$
$$(260)$$

$$l \tan \tfrac{1}{2} C = l \tan \tfrac{1}{2} (A + B) + l \cos \tfrac{1}{2} (a - b) - l \cos \tfrac{1}{2} (a + b).$$
$$(261)$$

64. From REIDT's Analogies (§ 44) we get the following equations :—

$$l \tan (45° - d''') = \tfrac{1}{2} \{ l \tan (s' + s'') + l \tan (s' - s'')$$
$$+ l \tan (d' + d'') - l \tan (d' - d'') \}. \qquad (262)$$

$$l \tan (45° - s''') = \tfrac{1}{2} \{ l \tan (s' + s'') - l \tan (s' - s'')$$
$$+ l \tan (d' + d'') + l \tan (d' - d'') \}. \qquad (263)$$

These formulae determine C and c when the angle B is acute. They possess the advantage of requiring only four logarithms instead of six, which are necessary if we calculate by the equations (258), (259). For the second triangle answering the given conditions, or for B obtuse, the formulae are—

$$l \tan s''' = \tfrac{1}{2} \{ l \tan (s' + s'') + l \tan (s' - s'') + l \tan (d' - d'')$$
$$- l \tan (d' + d'') \}. \qquad (264)$$

$$l \tan d''' + \tfrac{1}{2} \{ l \tan (s' - s'') + l \tan (d' - d'') + l \tan (d' + d'')$$
$$- l \tan (s' + s'') \}. \qquad (265)$$

The formulae (263)–(265) may be replaced by the following :—

$$l \tan (45° - s''') = l \tan (s' + s'') + l \tan (d' + d'') - l \tan (45° - d''').$$
(266)

$$l \tan d''' = l \tan (s' - s'') + l \tan (d' - d'') - l \tan s'''.$$ (267)

*Or thus :—*Let fall the perpendicular *CD*; then, denoting the arc *AD* by ϕ and the angle *ACD* by ψ, we have, from the right-angled triangle *ACD*,

$$\tan \phi = \tan b \cos A, \quad \tan \psi = \cot A / \cos b.$$

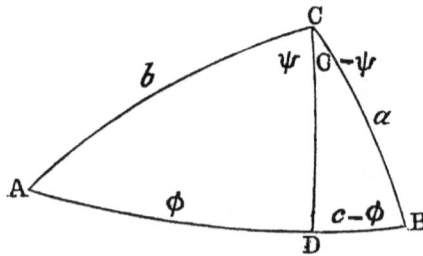

Fig. 19.

Then the sought parts are given by the equations

$$\sin B = \sin b \sin A / \sin a, \quad \cos (c - \phi) = \cos a \cos \phi / \cos b,$$

$$\cos (C - \psi) = \cot a \tan b \cos \psi.$$

Exercises.—XVIII.

1. Given $a = 73° 39' 38''$, $b = 120° 55' 35''$, $A = 88° 52' 42''$;
   find *B, C, c*.

2. „ $a = 150\ 57\ 5$, $b = 134\ 15\ 54$, $A = 144\ 22\ 42$;
   find *B, C. c*.

3. „ $a = 20\ 16\ 38$, $b = 56\ 19\ 40$, $A = 20\ 9\ 54$;
   find *B, C, c*.

65. *Given two angles A, B, and the side a opposite to one of them, to solve the triangle.*

The solution may be inferred at once from the reciprocal case, §§ 63, 64. In fact, the same equations solve both cases.

Third Pair of Cases.

66. *Being given the sides a, b, and the contained angle C, to find A, B, c.*

Napier's analogies give

$$l \tan \tfrac{1}{2}(A + B) = l \cot \tfrac{1}{2} C + l \cos \tfrac{1}{2}(a - b) - l \cos \tfrac{1}{2}(a + b). \tag{268}$$

$$l \tan \tfrac{1}{2}(A - B) = l \cot \tfrac{1}{2} C + l \sin \tfrac{1}{2}(a - b) - l \sin \tfrac{1}{2}(a + b). \tag{269}$$

These equations give $\tfrac{1}{2}(A + B)$ and $\tfrac{1}{2}(A - B)$; and therefore A and B, and then c, can be found from equation (47), or from (159) or (160).

EXAMPLE.—Given

$$a = 113° \, 2' \, 56'', \quad b = 82° \, 39' \, 28'', \quad C = 138° \, 50' \, 14''; \text{ find } A, B, c.$$

Type of the Calculation.

$\tfrac{1}{2}(a - b) = 15° \, 11' \, 14''$	$l \sin \tfrac{1}{2}(a - b)$	$= \bar{1}\cdot4184891$
$\tfrac{1}{2}(a + b) = 97 \quad 51 \quad 12$	$l \sin \tfrac{1}{2}(a + b)$	$= 1\cdot9959075$
$\tfrac{1}{2} C \quad = 69 \quad 25 \quad 7$	$l \cos \tfrac{1}{2}(a - b)$	$= \bar{1}\cdot9845438$
	$l \{-\cos \tfrac{1}{2}(a + b)\}$	$= \bar{1}\cdot1355722$
	$l \cot \tfrac{1}{2} C$	$= \bar{1}\cdot5746163$

Hence $l\{-\tan\tfrac{1}{2}(A+B)\} = \quad \cdot4235869$

$$\therefore \quad \tfrac{1}{2}(A + B) = 110° \, 39' \, 35'',$$

and $l \tan \tfrac{1}{2}(A - B) = \bar{2}\cdot9971969$;

$$\therefore \quad \tfrac{1}{2}(A - B) = 5° \, 40' \, 27''.$$

Hence $A = 116° \, 20' \, 2'', \quad B = 104° \, 59' \, 8''.$

To find C we have, from (159),

$$l \tan \tfrac{1}{2} c = l\{-\tan \tfrac{1}{2}(a+b)\} + l\{-\cos \tfrac{1}{2}(A+B)\} - l \cos \tfrac{1}{2}(A-B).$$

Now

$$l\{-\tan \tfrac{1}{2}(a+b)\} = \cdot 8603353, \quad l\{-\cos \tfrac{1}{2}(A+B)\} = \bar{1} \cdot 5475498,$$

$$l \cos \tfrac{1}{2}(A-B) = \bar{1} \cdot 9978668.$$

Hence $\quad l \tan \tfrac{1}{2} c = \cdot 4100083 ; \quad \therefore \; c = 137° \; 29' \; 3''.$

Observation.—In the foregoing calculation it is seen that, when an angle is between 90° and 180°, we have the sign minus before its cosine and its tangent, the reason of which is obvious.

Or thus :—Let fall the perpendicular BE; then denoting the arc AE by θ, CE will be $b - \theta$. Then, from the right-angled triangles, we have

$$\tan (b-\theta) = \tan a \cdot \cos C, \quad \tan A/\tan C = \sin (b-\theta)/\sin \theta.$$

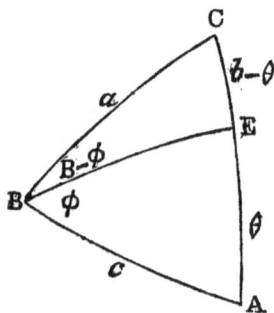

Fig. 20.

The first equation determines θ, and the second A. In a similar manner B may be found. Lastly, from the same triangles, we have $\cos c/\cos a = \cos \theta/\cos (b - \theta)$. Hence c is found.

EXERCISES.—XIX.

1. Given $a = 88° \; 12' \; 20''$, $\quad b = 124° \; 7' \; 17''$, $\quad C = 50° \; 2' \; 1''$;

   find A, B, c.

2. ,, $\quad a = 110 \; 55 \; 35$, $\quad b = 88 \; 12 \; 20$, $\quad C = 47 \; 42 \; 1$;

   find A, B, c.

3. ,, $\quad a = 65 \; 15 \; 12$, $\quad b = 47 \; 42 \; 1$, $\quad C = 59 \; 4 \; 25$;

   find A, B, c.

67. *Given two angles A, B, and the adjacent side c, to find*
a, b, C.

From Napier's Analogies we get

$$l \tan \tfrac{1}{2}(a+b) = l \tan \tfrac{1}{2}c + l \cos \tfrac{1}{2}(A-B) - l \cos \tfrac{1}{2}(A+B). \tag{270}$$

$$l \tan \tfrac{1}{2}(a-b) = l \tan \tfrac{1}{2}c + l \sin \tfrac{1}{2}(A-B) - l \sin \tfrac{1}{2}(A+B). \tag{271}$$

Hence a, b are known, and C can be found from (268) or (269).

Or thus :—Let fall the perpendicular BE (see last fig.) ; then
denoting the angle ABD by ϕ, the angle DBC will be $B - \phi$.
Then from the triangles ABD, CBD, we get

$$\cot \phi = \tan A \cos c, \quad \tan a = \cos \phi \tan C / \cos (B - \phi).$$

The first of these formulae determines ϕ, and the second a.
Similarly b may be found.

Again, from the same triangles, we have

$$\sin \phi : \sin (B - \phi) :: \cos A : \cos C.$$

Hence C is found.

68. The following simple and elementary methods of solving
the various cases of oblique-angled triangles, by dividing each
into two right-angled triangles, are due to CAUCHY.

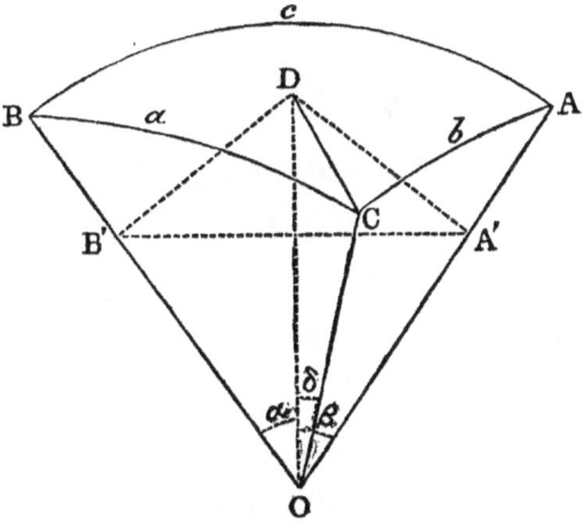

Fig. 21.

Let O be the centre of the sphere, ABC the spherical triangle.

Draw CD perpendicular to the plane AOB, also DA', DB' perpendicular to OA, OB. Then it is evident that CA', CB' are also perpendicular to OA, OB.

Let $$B'OD = a,\ DOA' = \beta,\ DOC = \delta;$$

then we have

$$CB' = \sin a,\ OB' = \cos a;\ CA' = \sin b,\ OA' = \cos b;$$

$$CD = \sin \delta,\ OD = \cos \delta;\ CA'D = A,\ CB'D = B.$$

The triangles $A'OD,\ B'OD,\ A'CD,\ B'CD$, have the angles $A'\ B'\ D$ right;

$$\therefore\ DO = \cos \delta = \frac{OB'}{\cos B'OD} = \frac{OA'}{\cos A'OD}.$$

Hence
$$\frac{\cos a}{\cos a} = \frac{\cos b}{\cos \beta}. \tag{272}$$

$$DC = \sin \delta = CB' \sin CB'D = CA' \sin CA'D.$$

Hence
$$\sin a . \sin B = \sin b \sin A. \tag{273}$$

$$DB' = OB' \tan B'OD = CB' \cos CB'D.$$

Hence
$$\tan a = \tan a \cos B. \tag{274}$$

$$DA' = OA' \tan A'OD = CA' \cos CA'D.$$

Hence
$$\tan \beta = \tan b \cos A. \tag{275}$$

From (272) we get

$$\tan \tfrac{1}{2}(a + b) \tan \tfrac{1}{2}(a - b) = \tan \tfrac{1}{2}(a + \beta) \tan \tfrac{1}{2}(a - \beta);$$

but
$$(a + \beta) = c. \tag{276}$$

Hence

$$\tan \tfrac{1}{2}(a + b) \tan \tfrac{1}{2}(a - b) = \tan \tfrac{1}{2}c . \tan \tfrac{1}{2}(a - \beta). \tag{277}$$

The formulae (272)–(277) solve all the cases of oblique-angled triangles.

1st Case.—*Given the three sides.*

(276) gives $(a + \beta)$,   (277) $(a - \beta)$,   (274), (275) give A, B.

2nd Case.—*Given the sides a, b, and the angle A.*

(273) gives B, (274), (275) give a, β, (276) gives c, and (273) C.

3rd Case.—*Given the angle A and the adjacent sides.*

(275) gives β, (276) a, (272) a, and (273) determines B and C.

EXERCISES.—XX.

1. Prove that

$$\cos \tfrac{1}{2}(a+b)\cos \tfrac{1}{2}(a-b)\tan \tfrac{1}{2}C = \sin a \cos B + \cos A \sin b. \quad (278)$$

2–7. Solve a right-angled triangle, being given—1°. c, $a+b$; 2°. c, $a-b$; 3°. a, $b+c$; 4°. a, $b-c$; 5°. c, $A-B$; 6°. c, p.

8. If $1 + \cos a + \cos b + \cos c = 0$, prove that each median is the supplement of the corresponding side.

9. In the same case, prove that the spherical excess is two right angles.

10. In the same case, prove that the arcs joining the middle points of two sides are each $= 90°$.

11–17. If $a+b+c=\pi$, prove—

1°. $\cos a = \tan \tfrac{1}{2}B \tan \tfrac{1}{2}C.$ 2°. $\sin^2 \dfrac{A}{2} = \cot b \cot c.$

3°. $\cos^2 \dfrac{A}{2} = \dfrac{\cos a}{\sin b \sin c}.$ 4°. $\tan^2 \dfrac{A}{2} = \dfrac{\cos b \cos c}{\cos a}.$

5°. $\sin^2 \dfrac{A}{2} + \sin^2 \dfrac{B}{2} + \sin^2 \dfrac{C}{2} = 1.$ 6°. $\cos A + \cos B + \cos C = 1.$

7°. $\operatorname{cosec}(A-E) + \operatorname{cosec}(B-E) + \operatorname{cosec}(C-E) = \operatorname{cosec} E.$

18. ABC is a spherical triangle right-angled at C; if with A, B as poles great circles $HFKL$, $DEFG$ be described, meeting the sides CA, CB, AB of the triangle in the pairs of points E, H; K, G; L, D, respectively; prove that the five triangles ABC, ADE, HEF, FGK, KLB have all the same circular parts.—(NAPIER.)

19–22.—Deduce from the analogies of DELAMBRE or NAPIER the following convergent series :—

1°. $\dfrac{a-b}{2} = \dfrac{c}{2} - \cot \dfrac{A}{2}\tan \dfrac{B}{2}\sin c + \tfrac{1}{2}\cot^2 \dfrac{A}{2}\tan^2 \dfrac{B}{2}\sin 2c - \&c.$

(BRÜNNOW).   (279)

2°. $\dfrac{c}{2} = \left(\dfrac{a-b}{2}\right) + \cot \dfrac{A}{2}\tan \dfrac{B}{2}\sin(a-b) + \tfrac{1}{2}\cot^2 \dfrac{A}{2}\tan^2 \dfrac{B}{2}\sin 2(a-b) + \&c.$

(*Ibid.*)   (280)

3°. $\dfrac{a+b}{2} = \dfrac{c}{2} + \tan \dfrac{A}{2}\tan \dfrac{B}{2}\sin c + \tfrac{1}{2}\tan^2 \dfrac{A}{2}\tan^2 \dfrac{B}{2}\sin 2c + \&c.$

(*Ibid.*)   (281)

4°. $\dfrac{c}{2} = \dfrac{a+b}{2} - \tan \dfrac{A}{2}\tan \dfrac{B}{2}\sin(a+b) + \tfrac{1}{2}\tan^2 \dfrac{A}{2}\tan^2 \dfrac{B}{2}\sin 2(a+b) - \&c.$

(*Ibid.*)   (282)

23. Prove

$$\cos 2A + \cos 2b - (\cos 2a + \cos 2B) = \cos 2A \cos 2b - \cos 2a \cos 2B.$$

$$(283)$$

24. Given any three of the six quantities s', s'', s''', d', d'', d''' of § 43, solve the triangle.

25–32. Solve a spherical triangle, being given—1°. A, B, $a+b$; 2°. A, B, $a-b$; 3°. C, c, $a+b$; 4°. a, B, $C+c$; 5°. c, b, h_c; 6°. a, B, E; 7°. A, b, $a+c$; 8°. $a\,b$, E, &c.

33. Prove

$$\tan (45° - s') \cot (45° = d') = \cot (s'' - s''') \tan (d'' - d'''). \quad (284)$$

34. Prove

$$\tan (45° - s') \tan (45° - d') = \tan (s'' + s''') \tan (d'' - d'''). \quad (285)$$

35. Prove

$$\tan \tfrac{1}{2} (B + C) + \tan \tfrac{1}{2} (B - C) = 2 \cot \tfrac{1}{2} A \sin b \div \sin (b + c). \quad (286)$$

36. If the cosines of the tangents drawn from any point P to two small circles have a given ratio, prove that the locus of P is a great circle.

37. In the same case, if the sum or the difference of the cosines of the tangents be given, prove that the locus of P is a circle.

38. State and prove the series similar to (279), (280) that may be obtained from the first and second of Napier's Analogies.

39. If c_1, c_2 be the values of the third side, when A, a, b are given, and the triangle is ambiguous, prove that

$$\tan \frac{c_1}{2} \tan \frac{c_2}{2} = \tan \tfrac{1}{2} (a + b) \tan \tfrac{1}{2} (a - b). \quad (287)$$

40. If A, B, C, &c., be the angular points of a regular polygon of n sides inscribed in a small circle, whose spherical centre is O and radius r; prove, if P be any point on the sphere, that

$$\Sigma \cos AP = n \cos r . \cos OP. \quad (288)$$

CHAPTER IV.

VARIOUS APPLICATIONS.

SECTION I.—THEORY OF TRANSVERSALS.

69. DEF. XVIII.—*Being given three points A, B, X on the same arc of a great circle, the ratio $\dfrac{\sin XA}{\sin XB}$ is called the ratio of section (AB, X).*

The arcs XA, XB are considered of the same or of different signs, according as they are measured in the same or in different directions from X. It is seen that it makes no difference whether we take for XB the arc $- XMB$ or $+ XAX'B$, these arcs having the same sign.

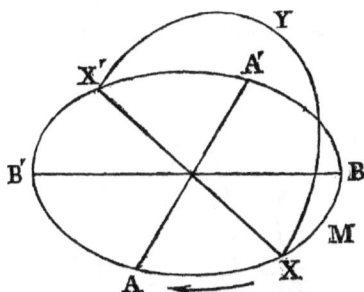

Fig. 22.

If A', B', X' be the antipodes of A, B, X, the ratio of section (AB, X) is positive if X be on BA' or AB',

negative if X be on AB or $A'B'$,

null   ,,   at A or A',

infinite   ,,   at B or B'.

We may remark that

$$(AB, X) = (AB, X') = (A'B', X) = - (A'B, X);$$

since it follows that when an arc AB meets another XYX', it is indifferent whether we take for the ratio (AB, X) or (AB, X').

DEF. XIX.—*If A, B, X, Y be four points on the same great circle, the ratio of the two ratios of section (AB, X) (AB, Y) or*

$$\frac{\sin XA}{\sin XB} : \frac{\sin YA}{\sin YB} \text{ is called the anharmonic ratio of the four points.}$$

If the ratio $= -1$, the points X, Y divide AB harmonically. For example, the two bisectors of an angle of a triangle divide the opposite side harmonically.

With four points on an arc of a great circle, the same as with four points on a right line, we can, as in *Sequel to Euclid*, p. 127, form six anharmonic ratios, any one of which may be called the anharmonic ratio of the points.

DEFINITION XX.—*When three arcs of great circles a, β, γ pass through the same point M, and are intersected in A, C, B by the great*

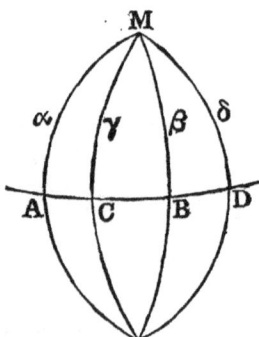

Fig. 23.

circle described, with M as pole, the ratio of section $(a\beta, \gamma)$

$$= \frac{\sin CA}{\sin CB} = \frac{\sin (\gamma a)}{\sin (\gamma \beta)}.$$

DEFINITION XXI.—*The anharmonic ratio of four great circles* (a, β, γ, δ) *passing through the same point M is* (αβγδ)

$$= \frac{sin(\gamma a)}{sin(\gamma\beta)} : \frac{sin(\delta a)}{sin(\delta\beta)}, \quad \text{or} \quad \frac{sin\ CA}{sin\ CB} : \frac{sin\ DA}{sin\ DB}.$$

If (αβγδ) = − 1, the pencil a, β, γ, δ is said to be harmonic.

70. If a great circle intersects the sides of a triangle *ABC* in the points *A'*, *B'*, *C'*, then,

$$1°. \quad (AB,\ C')(BC,\ A')(CA,\ B') = 1. \qquad (289)$$
$$2°. \quad (ab,\ CC')(bc,\ AA')(ca,\ BB') = 1. \qquad (290)$$

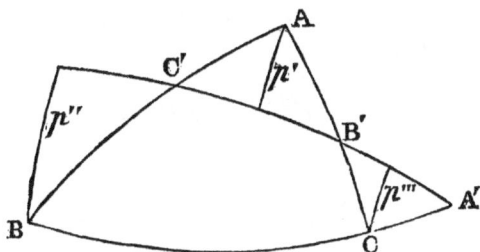

Fig. 24.

To prove 1°—If the sides *AB*, *CA* be cut internally, and *BC* externally, and perpendiculars p', p'', p''' be drawn to the transversal, we have by the properties of right-angled triangles,

$$(AB, C')= -\frac{sin\ p'}{sin\ p''},\ (BC, A')=\frac{sin\ p''}{sin\ p'''},\ \text{and}\ (CA, B') = -\frac{sin\ p'''}{sin\ p'}.$$

Hence, by multiplication, we have

$$(AB,\ C')(BC.\ A')(CA.\ B') = 1.$$

To prove 2°—We have, by equation (55), the equalities

$$(AB,\ C') = (ab,\ CC')\div\frac{sin\ A}{sin\ B};\ (BC, A') = (bc,\ AA')\div\frac{sin\ B}{sin\ C};$$
$$(CA,\ B') = (ca,\ BB')\div\frac{sin\ C}{sin\ A}.$$

Hence, by multiplication, the proposition is proved.

Reciprocally.—If the points A', B', C' satisfy either of these relations, they lie on a great circle.

Cor. 1.—The great circle which bisects two sides of a triangle meets the third side at the distance of 90° from its middle point.

Cor. 2.—The feet of two internal bisectors, and the foot of the third external bisector, of the angles of a triangle lie on the same great circle.

71. If the arcs which join the vertices of a triangle ABC to the same point O of the sphere meet the opposite sides in the points A', B', C', then,

$$1°. \quad (ab, CC')(bc, AA')(ca, BB') = -1. \qquad (291)$$

$$2°. \quad (AB, C')(BC, A')(CA, B') = -1. \qquad (292)$$

1°. This follows from applying the theorem, § 29, to the three triangles AOB, BOC, COA, and considering that if the point O be inside the triangle ABC, the three ratios of section (ab, CC'), &c., are negative; and if O be outside, two are positive and one negative.

2° follows from 1° by equation (55).

Reciprocally, if the points A', B', C' satisfy either of the equalities (291), (292), the arcs AA', BB', CC' are concurrent.

Cor. 1.—The three medians AA', BB', CC', are concurrent.

Cor. 2.—The three altitudes, h_a, h_b, h_c of a spherical triangle are concurrent.

Cor. 3.—The homologous sides of two supplemental triangles intersect in points situated on the same great circle, having as pole the common orthocentre of the two triangles.

Cor. 4.—The arcs which join the vertices of a spherical triangle to the points of contact of opposite sides with inscribed circle, or with any of the escribed, meet in the same point. (*The* Gergonne *point of the Triangle.*)

72. The anharmonic ratio of a pencil $(\alpha\beta\gamma\delta)$ of four great circles (see fig., Def. III.) is equal to the anharmonic ratio $(ABCD)$ of the four points in which it is intersected by a transversal.

DEM.—We have, equation (55),

$$(\alpha\beta, \gamma) = (AB, C) \cdot \frac{\sin A}{\sin B},$$

and

$$(\alpha\beta, \delta) = (AB, C) \cdot \frac{\sin A}{\sin B}.$$

Hence, by division, $(\alpha\beta\gamma\delta) = (ABCD).$ (293)

73. *Each diagonal of a complete spherical quadrilateral is divided harmonically by the two remaining diagonals.*

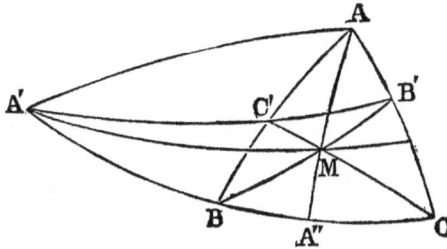

Fig. 25.

DEM.—Let the quadrilateral be $BCB'C'$; AA', BB', CC' its three diagonals. Let BB', CC' intersect in M. Join AM, and produce to cut BC in A''.

Now we have (Art. 70),

$$(AB, C')(BC, A'')(CA, B') = -1,$$

and (Art. 71),

$$(AB, C')(BC, A')(CA, B') = +1.$$

Hence $(BC, A'') = -(BC, A')$; $\therefore$ A', A'', B, C

are harmonic points. Therefore $(A - A'A''BC)$ is a harmonic pencil. Hence the proposition is proved.

Exercises.—XXI.

1–2. If the medians AA', BB', CC' of the triangle ABC intersect in G, prove—

$$1°. \quad \sin GA \div \sin GA' = 2 \cos \tfrac{1}{2} a. \tag{294}$$

$$2°. \quad \frac{\sin AA'}{\sin A'G} = \frac{\sin BB'}{\sin B'G} = \frac{\sin CC'}{\sin C'G}. \tag{295}$$

3. If through a fixed point P we draw any two transversals PAB, $PA'B'$ to a fixed angle XSY, the locus of the points of intersection of the arcs AB', $A'B$ is a great circle called the polar of P, with respect to the angle XSY.

4. If two spherical triangles ABC, $A'B'C'$ are such that the arcs AA', BB', CC' are concurrent, the pairs of corresponding sides AB, $A'B'$; BC, $B'C'$; CA, $C'A'$ intersect on the same great circle.

This may be proved by transversals (see *Sequel to Euclid*, p. 131), or by considering the tetrahedrons $(O - ABC)$, $(O - A'B'C')$ cut by the same plane, which gives two rectilineal triangles in perspective.

5. If we take on the three sides of a triangle from their middle points arcs equal to a quadrant, the six points thus obtained are on the same great circle.

6. If the arcs AP, BP, CP meet the sides BC, CA, AB in A', B', C'; and if A'', B'', C'' be the symmetriques* of A', B', C', with respect to the middle points of the sides, then AA'', BB'', CC'' meet in the same point P', called the isotomic conjugate of P, with respect to the triangle.

7. Prove that the three arcs AD, BE, CF, each bisecting the area of a spherical triangle ABC, are concurrent.—(Steiner.)

From the given conditions the spherical excess of each of the triangles BAD, CAD is E. Hence

$$\sin \tfrac{1}{2} AD = \sqrt{\frac{\sin \tfrac{1}{2} E . \sin (B - \tfrac{1}{2} E)}{\sin BAD . \sin ADB}} = \sqrt{\frac{\sin \tfrac{1}{2} E . \sin (C - \tfrac{1}{2} E)}{\sin CAD . \sin ADC}} ;$$

$$\therefore \quad \sin BAD : \sin CAD : : \sin (B - \tfrac{1}{2} E) : \sin (C - \tfrac{1}{2} E),$$

from which and two similar proportions the proposition follows.

---

* For shortness, we say that the extremities of an arc of a great circle are symmetriques, with respect to the middle of that arc.

DEF. XXII.—*We shall call the normal co-ordinates of a point M, with respect to a triangle ABC, quantities proportional to the sines of arcs drawn from M perpendicular to the sides of the triangle, and denote them by* δ_a, δ_b, δ_c.

DEF. XXIII.—*We shall call the triangular co-ordinates of M half the products of the sines of the perpendiculars from M on the sides, multiplied by the sines of the sides.* The triangular co-ordinates of M are equal to the *Staudtians* of the triangles AMB, BMC, CMA; we shall denote them by n_a, n_b, n_c.

8. If arcs AM, BM, CM meet the sides of ABC in A', B', C', respectively, prove that

$$(BC, A') = n_c : n_b ; \quad (CA, B') = n_a : n_c ; \quad (AB, C') = n_b : n_a.$$

9. If two points be isotomic conjugates, they have reciprocal triangular co-ordinates.

10. If three arcs drawn through A, B, C be the symmetriques of any three arcs AM, BM, CM, with respect to the bisectors of the angles A, B, C, they meet in a common point M', called the isogonal conjugate of M.

11. If two points be isogonal conjugates with respect to a triangle, their normal co-ordinates are reciprocals.

12. If a transversal T cuts the sides of ABC in A', B', C', the symmetriques of A', B', C', with respect to the middle points of BC, CA, AB, are upon the same arc of a great circle T, called the *isotomic transversal* of T.

13. In the same case, the symmetriques of the arcs AA', BB', CC', with respect to the bisectors of the angles A, B, C, meet the sides of ABC in points which lie on the same great circle T'', called the *isogonal transversal* of T.

14. Prove that the triangular co-ordinates of G, the point of intersection of the medians of a triangle, are equal to one another.

15. If A_1, B_1, C_1 be the harmonic conjugates of the points A', B', C', in which a transversal T cuts the sides of ABC with respect to the sides, then the arcs AA_1, BB_1, CC_1 co-intersect in a point τ, called the *trilinear pole* of T.

T is called the trilinear polar of τ.

16. If G be the intersection of the medians, M any point of the sphere, n' the Staudtian of the triangle BGC; then

$$\frac{\cos MA + \cos MB + \cos MC}{\cos MG} = \text{constant} = n \div n'. \qquad (296)$$

Dᴇᴍ.— $\qquad \cos MB + \cos MC = 2 \cos \dfrac{a}{2} \cos MA'.$

$$\cos MA' . \sin AG + \cos MA . \sin GA' = \cos MG . \sin AA' ;$$

$$\therefore \quad \cos MB + \cos MC = 2 \cos \frac{a}{2} \left\{ \frac{\cos MG . \sin AA'}{\sin AG} - \frac{\cos MA . \sin GA}{\sin AG} \right\} ;$$

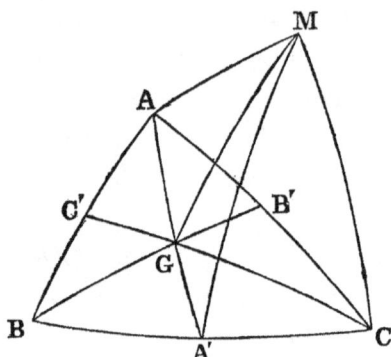

Fig. 26.

$$\therefore \quad \frac{\cos MA + \cos MB + \cos MC}{\cos MG} = 2 \cos \frac{a}{2} . \frac{\sin AA'}{\sin AG},$$

since $\qquad\qquad 2 \cos \dfrac{a}{2} = \dfrac{\sin AG}{\sin GA'}.$

Hence $\qquad \dfrac{\cos MA + \cos MB + \cos MC}{\cos MG} = \dfrac{\sin AA'}{\sin GA'} = \dfrac{n}{n'}.$

17. Calculate the norm of the sides of BGC.

We have

$$\frac{n'}{n} = \frac{\sin GA'}{\sin AA'}, \quad 2 \cos \frac{a}{2} = \frac{\sin (AA' - GA')}{\sin GA'}, \quad \cos b + \cos c = 2 \cos \frac{a}{2} . \cos AA'.$$

We shall eliminate GA' and AA' between these equations. The second gives

$$2 \cos \frac{a}{2} \sin GA' = \sin AA' \cos GA - \cos AA' \sin GA' ;$$

$$\therefore \quad \sin GA' \left(2 \cos \frac{a}{2} + \cos AA' \right) = \cos GA' . \sin AA' ;$$

$$\therefore \quad \frac{n'}{n} \left(2 \cos \frac{a}{2} + \cos AA' \right) = \cos GA' = \sqrt{1 - \frac{n'^2}{n^2} \sin^2 AA'},$$

$$\frac{n'^2}{n^2} = \frac{1}{4 \cos^2 \dfrac{a}{2} + 4 \cos \dfrac{a}{2} \cos AA' + 1} = \frac{1}{4 \cos^2 \dfrac{a}{2} + 2 (\cos b + \cos c) + 1}.$$

Hence $$\frac{n'}{n} = \frac{1}{\sqrt{1 \pm 2(1 + \cos a + \cos b + \cos c)}}. \qquad (297)$$

18. If tangents be drawn at A, B, C to the circumcircle of the triangle ABC, forming a triangle $A'B'C'$, the arcs AA', BB', CC', are concurrent. The point of concurrence, K, is called the LEMOINE *point of the Triangle.*

19. Prove that the normal co-ordinates of K are

$$\sin (A - E), \ \sin (B - E), \ \sin (C - E).$$

20. The triangular co-ordinates of the orthocentre are $\tan A$, $\tan B$, $\tan C$, and the normal co-ordinates, $\sec A$, $\sec B$, $\sec C$.

DEF. XXIV.—*The isogonal conjugate of O, the intersection of the medians, is called the* SYMMEDIAN *point.*

21. Prove that the Symmedian point of a spherical triangle does not coincide with its Lemoine point. Its normal co-ordinates are $\sin a$, $\sin b$, $\sin c$.

22. The normal co-ordinates of the pole of the circumcircle are

$$\cos (A - E), \quad \cos (B - E), \quad \cos (C - E).$$

23. If M be any point of the sphere, and the arcs MA, MB, MC meet the sides BC, CA, AB in A', B', C', if O be the pole of the circumcircle,

$$\frac{\sin MA'}{\sin AA'} + \frac{\sin MB'}{\sin BB'} + \frac{\sin MC'}{\sin CC'} = \frac{\cos MO}{\cos R}. \quad \text{(STEINER.)} \quad (298)$$

24. If two equianharmonic pencils have a common ray, the intersection of three corresponding pairs of rays lie on a great circle. Compare *Sequel to Euclid*, Prop. v., p. 131.

SECTION II.—INCIRCLES.

74. *To find the radius of the incircle of a spherical triangle ABC.*

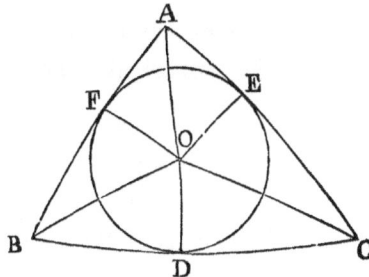

Fig. 27.

Sol.—Bisect the angles A, B by the arcs AO, BO. O is the incentre required; and the perpendiculars *OD, OE, OF* on the sides are the angular radii.

DEM.—It is easy to see that *OD, OE, OF* are all equal; also that $AF = s - a$. Now from the right-angled triangle *OAF* we have, equation (110),

$$\tan OAF = \tan OF \div \sin AF;$$

or, denoting the radius by *r*,

$$\tan \tfrac{1}{2} A = \tan r \div \sin (s - a). \tag{299}$$

Hence, substituting for $\tan \tfrac{1}{2} A$ its value, equation (22), we get

$$\tan r = \sqrt{\frac{\sin(s-a)\,\sin(s-b)\,\sin(s-c)}{\sin s}} = \frac{n}{\sin s}. \tag{300}$$

Cor. 1.—If in the expression for tan *r* we substitute for *b, c* $(\pi - b)$, $(\pi - c)$, we get the expression for the in-radius of the colunar triangle *BCA'* formed by producing the sides *AB, AC*. Hence, denoting it by r_a, we get

$$\tan r_a = \frac{n}{\sin (s - a)}. \tag{301}$$

Similarly,

$$\tan r_b = \frac{n}{\sin (s - b)}, \tag{302}$$

and

$$\tan r_c = \frac{n}{\sin (s - c)}. \tag{303}$$

Cor. 2.—From the equations (299), (300) we get the following formulae for solving a spherical triangle when the three sides are given :—

$$l \tan r = \tfrac{1}{2} \{ l \sin (s - a) + l \sin (s - b) + l \sin (s - c) - l \sin s \}. \tag{304}$$

$$l \tan \tfrac{1}{2} A = l \tan r - l \sin (s - a), \&c. \tag{305}$$

DEF. XXV.—*The incircles of the colunar triangles are called escribed circles.*

Exercises.—XXII.

1. Prove $\tan r = \sin a \sin \tfrac{1}{2} B \sin \tfrac{1}{2} C \sec \tfrac{1}{2} A,$ (306)

2.   ,, $\tan r = N \div 2 \cos \tfrac{1}{2} A \cos \tfrac{1}{2} B \cos \tfrac{1}{2} C.$ (307)

3.   ,, $\cot r = \dfrac{1}{2N} \{ \sin(A - E) + \sin(B - E) + \sin(C - E) - \sin E \}.$

                                                         (308)

4.   ,,   if $a + b + c = \pi$, prove $\tan r = \tan \tfrac{1}{2} A \tan \tfrac{1}{2} B \tan \tfrac{1}{2} C.$ (309)

5.   ,, $\tan r \cdot \tan r_a \cdot \tan r_b \tan r_c = n^2.$ (310)

6.   ,, $\tan r_a = \sin a \cos \tfrac{1}{2} B \cos \tfrac{1}{2} C \sec \tfrac{1}{2} A.$ (311)

7.   ,, $\cot r_a = \dfrac{1}{2N} \{ \sin E + \sin(B - E) + \sin(C - E) - \sin(A - E) \}.$

                                                         (312)

8.   ,, $\cot r - \cot r_a = \dfrac{1}{N} \{ \sin(A - E) - \sin E \}.$ (313)

9. Prove that the centre of the incircle is the orthocentre of the triangle formed by the excentres.

10. Prove $\cot r_a + \cot r_b + \cot r_c = (\cot \tfrac{1}{2} A + \cot \tfrac{1}{2} B + \cot \tfrac{1}{2} C) \div \sin s.$

                                                             (314)

11. Prove that the common tangents of the escribed circles taken in pairs are $a + b$, $b + c$, $c + a$, respectively.

12. If O_a, O_b, O_c be the centres of the escribed circles, prove that

$$\cos OO_a : \cos OO_b : \cos OO_c : : \frac{\cos r_a}{\cos(s - a)} : \frac{\cos r_b}{\cos(s - b)} : \frac{\cos r_c}{\cos(s - c)}. \quad (315)$$

13. Prove that $\cot r + \cot r_a + \cot r_b + \cot r_c$

$$= \frac{1}{N} \{ \sin E + \sin(A - E) + \sin(B - E) + \sin(C - E) \}. \quad (316)$$

14. Prove

$$\sin^2 AO : \sin^2 BO : \sin^2 CO : : \frac{\sin(s - a)}{\sin a} : \frac{\sin(s - b)}{\sin b} : \frac{\sin(s - c)}{\sin c}. \quad (317)$$

15. Prove that the cosines of the angles of the triangle O_a, O_b, O_c are respectively equal to

$$\cos s \cdot \sin \tfrac{1}{2} A, \quad \cos s \cdot \sin \tfrac{1}{2} B, \quad \cos s \cdot \sin \tfrac{1}{2} C.$$

Section III.—Circumcircles.

75. *To find the circumradius of a spherical triangle ABC.*

Sol.—Bisect the arcs BC, CA at D, E, and let O be the intersection of perpendiculars to BC, CA, at D, E; then O is the circumcentre.

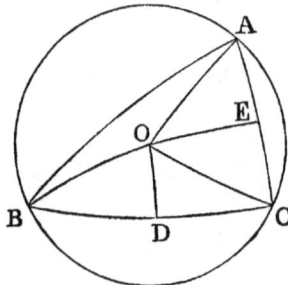

Fig. 28.

Dem.—Join OA, OB, OC; then, equation (113), cos OB = cos BD. cos OD, and cos OC = cos DC. cos OD. Hence OB = OC. Similarly, $OC = OA$; therefore O is the circumcentre of the triangle ABC. Again, the angle

$$OAB = OBA, \quad OBC = OCB, \text{ and } OCA = OAC;$$
$$\therefore \quad OCB + A = \tfrac{1}{2}(A + B + C) = S;$$
$$\therefore \quad OCB = S - A = 90° - (A - E).$$

Let $OC = R$; then, from the triangle OCD we have, equation (109),

$$\cos OCD = \tan DC \div \tan OC = \tan \tfrac{1}{2} a \div \tan R;$$
$$\therefore \quad \tan R = \tan \tfrac{1}{2} a \div \sin (A - E); \quad\quad (318)$$

and, substituting for $\tan \tfrac{1}{2} a$ its value from equation (86), we get

$$\cot R = \sqrt{\frac{\sin (A - E) \sin (B - E) \sin (C - E)}{\sin E}} = \frac{N}{\sin E}.$$

$$(319)$$

Cor. 1.—If in equation (319) we substitute for B, C, $\pi - B$, $\pi - C$, respectively, we get the circumradius of the colunar triangle BCA'. Hence, denoting it by R_A, we have

$$\cot R_A = \frac{N}{\sin (A - E)}. \tag{320}$$

Similarly,

$$\cot R_B = \frac{N}{\sin (B - E)}, \tag{321}$$

and

$$\cot R_C = \frac{N}{\sin (C - E)}. \tag{322}$$

Cor. 2.—From the equations (318), (319), we get the following formulae for the solution of spherical triangles, when the angles are given. Thus—

$$l \cot R = \tfrac{1}{2} \left\{ l \sin(A - E) + l \sin (B - E) + l \sin (C - E) - l \sin E \right\}. \tag{323}$$

$$l \cot \tfrac{1}{2} a = l \cot R - l \sin (A - E), \text{ \&c.} \tag{324}$$

Exercises.—XXIII.

1. Prove
$$\tan R = \frac{\sin \tfrac{1}{2} a}{\sin A \cos \tfrac{1}{2} b \cos \tfrac{1}{2} c}. \tag{325}$$

2. ,,
$$\tan R = \frac{2 \sin \tfrac{1}{2} a \sin \tfrac{1}{2} b \sin \tfrac{1}{2} c}{n}. \tag{326}$$

3. ,,
$$\tan R_A = \frac{\sin \tfrac{1}{2} a}{\sin A . \sin \tfrac{1}{2} b . \sin \tfrac{1}{2} c}. \tag{327}$$

4. ,,
$$\tan R_A = \frac{2 \sin \tfrac{1}{2} a \cos \tfrac{1}{2} b \cos \tfrac{1}{2} c}{n}. \tag{328}$$

5. ,,
$$\tan R - \tan R_A = \frac{1}{n} \left\{ \sin s - \sin (s - a) \right\}. \tag{329}$$

6. ,,
$$\tan R . \tan R_A . \operatorname{tau} R_B . \tan R_C = \frac{1}{N^2}. \tag{330}$$

7. ,,
$$\tan R + \tan R_A = \cot r_b + \cot r_c. \tag{331}$$

8. ,,
$$\tan R_B + \tan R_C = \cot r + \cot r_a. \tag{332}$$

9. ,,
$$\tan R + \cot r = \tfrac{1}{2} \left\{ \tan R + \tan R_A + \tan R_B + \tan R_C \right\}. \tag{333}$$

10. Prove $\tan R \cdot \tan R_A + \tan R_B \cdot \tan R_C = \cot r \cdot \cot r_a + \cot r_b \cdot \cot r_c.$

$$(334)$$

11. „ $(\cot r + \tan R)^2 + 1 = \left(\dfrac{\sin a + \sin b + \sin c}{2n}\right)^2.$ $\quad(335)$

12. „ $(\cot r + \tan R)^2 + 1 = \left(\dfrac{\sin b + \sin c - \sin a}{2n}\right)^2.$ $\quad(336)$

13. „ $\tan^2 R + \tan^2 R_A + \tan^2 R_B + \tan^2 R_C$

$$= \cot^2 r + \cot^2 r_a + \cot^2 r_b + \cot^2 r_c. \quad (337)$$

14. Prove that the angles of intersection of the circumcircle of a spherical triangle with the circumcircles of the colunar triangles are equal to the angles of the triangle.

15. The angles of intersection of the sides of a spherical triangle with its circumcircle are $(A - E)$, $(B - E)$, $(C - E)$, respectively.

16. The angles of intersection of the circumcircles of the colunar triangles, in pairs, are equal to $A + B$, $B + C$, $C + A$, respectively.

17. If ABC be a triangle, right-angled at C, if the point C and the circumcircle of the triangle ABC be given in position, prove that the locus of the circumcentre of its colunar triangle ABC' is a great circle.

18. If δ be the spherical distance between the poles of the incircle and circumcircle of a spherical triangle, prove that

$$\cos^2 \delta - \cos^2 R = \cos^2 (R - r) - \cos^2 R \cos^2 r. \quad (338)$$

19. If δ_a denote the distance between the circumcentre and the incentre of BCA', prove

$$\cos^2 \delta_a - \cos^2 R = \cos^2 (R + r_a) - \cos^2 R \cos^2 r_a. \quad (339)$$

20. Prove $\cos \delta = \sin r \sin R \left(\dfrac{\sin a + \sin b + \sin c}{4 \sin \frac{1}{2}a \cdot \sin \frac{1}{2}b \cdot \sin \frac{1}{2}c}\right).$ $\quad(340)$

21. Prove $\cot r_a + \cot r_b + \cot r_c - \cot r = 2 \tan R.$ $\quad(341)$

22. In an equilateral spherical triangle, $\tan R = 2 \tan r.$

23. Prove that $\tan R \sin h_a = \dfrac{2 \sin \frac{1}{2}b \cdot \sin \frac{1}{2}c}{\cos \frac{1}{2}a}.$ $\quad(342)$

24. Prove that the *Lhuilierian* of a spherical triangle is equal to the *Lhuilierian* of each of its colunar triangles, or to that of its polar triangle, or any of the colunar triangles of the polar triangle.

25. If a, β, γ, δ denote the perpendiculars from any point in a small circle on the sides of an inscribed quadrilateral, whose lengths are a, b, c, d; prove that

$$\sin a \sin \gamma \cos \tfrac{1}{2}a \cos \tfrac{1}{2}c = \sin \beta \sin \delta \cos \tfrac{1}{2}b \cos \tfrac{1}{2}d. \quad (343)$$

26. In a quadrantal triangle, of which the side c is the quadrant, prove that

$$\cot R = \sin(C - E), \quad \cot R_A = \sin(B - E), \quad \cot R_B = \sin(A - E),$$
$$\cot R_C = \sin E. \tag{344}$$

27. If the angle A of a triangle remains constant, and also the perimeter, the envelope of the side BC is a small circle.

28. If the angle A remains constant, and also $b + c - a$, the envelope of BC is a small circle.

29–31. Construct and resolve a spherical triangle, being given

$$1°. \quad A, \ a, \ b + c; \quad 2°. \quad A, \ a, \ r; \quad 3°. \quad A, \ a, \ R.$$

32. Prove $\cos^2 R = \dfrac{\tan(A - E)\tan(B - E)\tan(C - E)}{\tan(A - E) + \tan(B - E) + \tan C - E)}.$ $\tag{345}$

33. Find the simplest formulae for r, r_a, r_b, r_c; R, R_A, R_B, R_C, for a diametral triangle; that is, for a triangle for which

$$C = A + B, \ \text{or} \ \sin^2 \frac{c}{2} = \sin^2 \frac{a}{2} + \sin^2 \frac{b}{2}.$$

34. If a spherical quadrilateral be such that it is inscribed in a small circle of radius R, and circumscribed to another of radius r, prove that if δ be the distance between the poles,

$$\sin(R + r + \delta)\sin(R + r - \delta)\sin(R - r + \delta)\sin(R - r - \delta) = \sin^4 r \cos^4 R.$$
$$\text{(Steiner.)} \quad (346)$$

Section IV.—Spherical Mean Centres.

Def. XXVI.—*If the triangular co-ordinates of a point M with respect to a triangle ABC be n_a, n_b, n_c, we have seen* (Ex. xxi., 8) *that the arcs AM, BM, CM divide BC, CA, AB in the spherical ratios $n_c : n_b$, $n_a : n_c$, $n_b : n_a$. M is called the spherical mean centre of the points with respect to the system of multiples n_a, n_b, n_c.*

76. If M be the mean centre of the points A, B, C for the multiples n_a, n_b, n_c, and P be any other point, then

$$n_a \cos AP + n_b \cos BP + n_c \cos CP = n \cdot \cos MP. \tag{347}$$

Dem.—We have by *Stewart's Theorem* (Euc. III. 17), from the triangle BPC,

$$\cos PB \sin A'C + \cos PC \sin BA' = \cos PA' \sin a;$$

and from $A'PA$,

$$\cos PA' \sin AM + \cos PA \sin MA' = \cos PM \sin AA'.$$

Eliminating PA', we get

$$\cos PB \sin A'C + \cos PC \sin BA' + \frac{\cos PA \sin a \sin MA'}{\sin AM}$$

$$= \frac{\cos PM \sin AA' \sin a}{\sin AM};$$

but

$$2n_a = \sin AM \sin BA' \sin a; \quad 2n_b = \sin AM \sin CA' \sin a;$$

$$2n_c = \sin a \sin MA' \sin a; \quad 2n = \sin a \sin AA' \sin a.$$

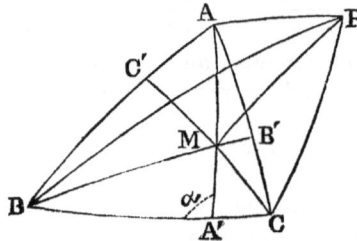

Fig. 29.

Hence, eliminating BA', CA', MA', AA', we get

$$n_a \cos PA + n_b \cos PB + n_c PC = n \cos PM.$$

Cor. 1.—If P be a point, such that for given multiples l, m, n, $l \cos AP + m \cos BP + n \cos CP$ is constant, the locus of P is a small circle.

Cor. 2.—If there be any number of fixed points A, B, C, &c., and a corresponding system of multiples l, m, n, &c., and P any point satisfying the condition $\Sigma (l \cos AP) =$ constant, the locus of P is a circle.

Cor. 3.—If ρ be the spherical radius of the circle, touching the inscribed and escribed circles of a spherical triangle (HART's circle),

$$\tan \rho = \tfrac{1}{2} \tan R. \tag{348}$$

DEM.—Let P be the pole of Hart's circle, I, I_a, I_b, I_c the poles of the in- and circumcircles; thus

$$IH = \rho - r, \quad I_a H = \rho + r_a, \quad I_b H = \rho + r_b, \quad I_c H = \rho + r_c.$$

Then it is evident that the staudtians of the triangles

$$I_a I_b I_c, \quad II_b I_c, \quad II_c I_a, \quad II_a I_b,$$

are proportional to

$$\frac{1}{\sin r}, \quad \frac{1}{\sin r_a}, \quad \frac{1}{\sin r_b}, \quad \frac{1}{\sin r_c}.$$

Hence, from equation (347), we have

$$\frac{\cos(\rho + r_a)}{\sin r_a} + \frac{\cos(\rho + r_b)}{\sin r_b} + \frac{\cos(\rho + r_c)}{\sin r_c} = \frac{\cos(\rho - r)}{\sin r};$$

$$\therefore \quad 4 \tan \rho = \cot r_a + \cot r_b + \cot r_c - \cot r = 2 \tan R.$$

Hence $\qquad\qquad \tan \rho = \tfrac{1}{2} \tan R.$

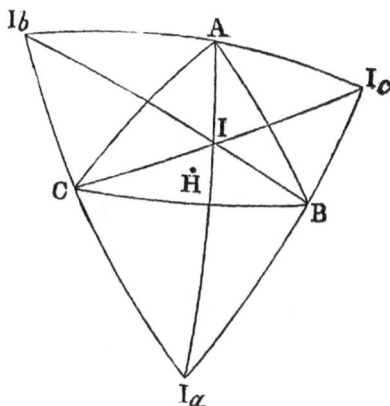

Fig. 30.

Cor. 4.—If p_a, p_b, p_c be the distances from A, B, C to a great circle T passing through M, then

$$n_a \sin p_a + n_b \sin p_b + n_c \sin p_c = 0. \qquad (349)$$

This follows from (347) by supposing P to be the pole of T.

Cor. 5.—If T be any great circle, and if $n_a \sin p_a + n_b \sin p_b + n_c \sin p_c = $ constant, the envelope of T is a small circle.

Cor. 6.—If A, B, C, &c., be any system of fixed points, and l, m, n, &c., a system of multiples, and if p_a, p_b, p_c, &c., be the perpendiculars from A, B, C, &c., on any great circle T; then, if $\Sigma(l \sin p_a)$ = constant, the envelope of T is a small circle.

Exercises.—XXIV.

1. If from the points I_a, I_b, I_c perpendiculars be drawn to the sides BC, CA, AB, respectively, these perpendiculars are concurrent.

2. If H be the pole of Hart's Circle, prove that

$$\cos AH \div \cos \rho = \cos \tfrac{1}{2} b \cos \tfrac{1}{2} c \div \cos \tfrac{1}{2} a. \qquad (350)$$

3. The arcs of connection of the vertices of a triangle to the points of contact of the opposite sides, with the circles inscribed in the corresponding colunar triangle, meet in the same point, called the Nagel *point of the triangle.*

4. If t_1, t_2, t_3 be the tgents from A, B, C to Hart's Circle, prove

$$\cos t_1 \cos t_2 \cos t_3 = \cos \frac{a}{2} \cos \frac{b}{2} \cos \frac{c}{2}.$$

5. The normal co-ordinates of the pole of Hart's Circle are

$$\cos (B - C), \quad \cos (C - A), \quad \cos (A - B).$$

6. The normal co-ordinates of Nagel's Point are

$$\sin^2 \tfrac{1}{2} A, \quad \sin^2 \tfrac{1}{2} B, \quad \sin^2 \tfrac{1}{2} C.$$

7-10 Prove the following relations :—

1°. $\sin (B + C)/\sin A = (\cos b + \cos c)/(1 + \cos a)$.

2°. $\sin (B - C)/\sin A = (\cos c - \cos b)/(1 - \cos a)$.

3°. $\dfrac{\sin (B + C) \cot \tfrac{1}{2} a}{\cos b + \cos c} = \dfrac{(\cos B + \cos C) \cot \tfrac{1}{2} A}{\sin (b + c)}.$

4°. $\dfrac{\sin (B + C) \tan \tfrac{1}{2} a}{\cos b - \cos c} = \dfrac{(\cos C - \cos B) \tan \tfrac{1}{2} A}{\sin (b + c)}.$

CHAPTER V.

SPHERICAL EXCESS.

77. In the preceding chapters we have made frequent use of the function of the angles of a triangle, called the *spherical excess*. In this chapter we shall enter into further detail, and give a more systematic account of its theory than could have been conveniently given in those chapters.

SECTION I.—FORMULAE RELATIVE TO E.

78. *Lemma.*—If the triangle BCA' be colunar with ABC, it has two sides equal to $\pi - b$, $\pi - c$, and their included angle is equal to A; therefore if $2E$ be the spherical excess of ABC, $2A - 2E$ will be the spherical excess of $A'BC$. Hence we have the following rule of transformation:—

RULE.—*In any formula containing the elements b, c, A, E of a spherical triangle, we may change the sides b, c into their supplement, and E into $A - E$.*

This rule supplies easy proofs of several propositions.

79. Let ABC be a spherical triangle; bisect BC, AC in

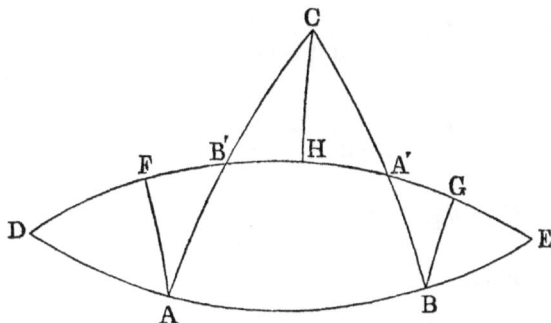

Fig. 31.

A', B'. Join $A'B'$, and produce to meet AB in D, E; let fall the perpendiculars AF, BG, CH. Then evidently the three pairs of triangles $AB'F$, $CB'H$; $A'BG$, $A'CH$; ADF, BEG are, two by two, equal in every respect. Hence it is easy to see that the angle $BAF = \frac{1}{2}(A + B + C)$; that is, $= 90° + E$; therefore $DAF = 90° - E$; also DF is the complement of $A'B'$, and DA of half AB; that is, $DA = 90° - \dfrac{c}{2}$.

80. Cagnoli's Theorem.—From the values of $\cos \frac{1}{2} a$, $\sin \frac{1}{2} b$, $\sin \frac{1}{2} c$ (§§ 32, 33), we get

$$\sin E = \frac{\sin \frac{1}{2} b \, \sin \frac{1}{2} c \, \sin A}{\cos \frac{1}{2} a};$$

but

$$\sin A = \frac{n}{\sin b \, \sin c}.$$

Hence

$$\sin E = \frac{n}{2 \cos \frac{1}{2} a \, \cos \frac{1}{2} b \, \cos \frac{1}{2} c}. \qquad (351)$$

This is Cagnoli's Theorem.

81. By the transformation of § 78, we get

$$\sin (A - E) = \frac{n}{2 \cos \frac{1}{2} a \, \sin \frac{1}{2} b \, \sin \frac{1}{2} c}. \qquad (352)$$

Hence, by interchange of letters,

$$\sin (B - E) = \frac{n}{2 \sin \frac{1}{2} a \, \cos \frac{1}{2} b \, \sin \frac{1}{2} c}. \qquad (353)$$

$$\sin (C - E) = \frac{n}{2 \sin \frac{1}{2} a \, \sin \frac{1}{2} b \, \cos \frac{1}{2} c}. \qquad (354)$$

82. *To find the value of* $\cos E$.

From the triangle DAF, § 79, we get

$$\sin DAF = \sin FD \div \sin AD,$$

or

$$\cos E = \frac{\cos A'B'}{\cos \frac{1}{2} c}; \qquad (355)$$

but　　　$\cos A'B' = \cos \tfrac{1}{2} a \cos \tfrac{1}{2} b + \sin \tfrac{1}{2} a \sin \tfrac{1}{2} b \cos C,$

from the triangle $A'B'C$.

Hence

$$\cos E = \frac{\cos \tfrac{1}{2} a \cos \tfrac{1}{2} b + \sin \tfrac{1}{2} a \sin \tfrac{1}{2} b \cos C}{\cos \tfrac{1}{2} c}. \qquad (356)$$

And substituting for $\cos C$ its value from equation (15), and reducing, we get

$$\cos E = \frac{\cos^2 \tfrac{1}{2} a + \cos^2 \tfrac{1}{2} b + \cos^2 \tfrac{1}{2} c - 1}{2 \cos \tfrac{1}{2} a \cos \tfrac{1}{2} b \cos \tfrac{1}{2} c}. \qquad (357)$$

Or thus :　　　$\cos b + \cos c = 2 \cos AA' \cos \dfrac{a}{2},$

$$\cos AA' + \cos \frac{a}{2} = 2 \cos A'B' \cos \frac{b}{2}.$$

Hence, eliminating $\cos AA'$, we get

$$\cos A'B' = \frac{2 \cos^2 \dfrac{a}{2} + \cos b + \cos c}{4 \cos \dfrac{a}{2} \cos \dfrac{b}{2}} = \frac{1 + \cos a + \cos b + \cos c}{4 \cos \dfrac{a}{2} \cos \dfrac{b}{2}}.$$

Hence from (355),　　　　　　　　　　　　　　　(358)

$$\cos E = \frac{1 + \cos a + \cos b + \cos c}{4 \cos \dfrac{a}{2} \cos \dfrac{b}{2} \cos \dfrac{c}{2}}. \qquad (359)$$

Cor. 1.—From (351) and (354) we get

$$\cot E = \frac{\cot \tfrac{1}{2} a \cot \tfrac{1}{2} b}{\sin C} + \cot C; \qquad (360)$$

and by interchange of letters we get

$$\cot E = \frac{\cot \tfrac{1}{2} b \cot \tfrac{1}{2} c}{\sin A} + \cot A. \qquad (361)$$

and

$$\cot E = \frac{\cot \tfrac{1}{2} c \cot \tfrac{1}{2} a}{\sin B} + \cot B. \qquad (362)$$

Cor. 2.—If the area of a spherical triangle and one of its angles be given, the product of the semitangents of the containing sides is given.

83. By the transformation of § 78 we get, from (357),

$$\cos(A - E) = \frac{\cos^2 \tfrac{1}{2} a + \sin^2 \tfrac{1}{2} b + \sin^2 \tfrac{1}{2} c - 1}{2 \cos \dfrac{a}{2} \sin \dfrac{b}{2} \sin \dfrac{c}{2}},$$

or $\quad \cos(A - E) = \dfrac{\sin^2 \tfrac{1}{2} b + \sin^2 \tfrac{1}{2} c - \sin^2 \tfrac{1}{2} a}{2 \cos \tfrac{1}{2} a \sin \tfrac{1}{2} b \sin \tfrac{1}{2} c}.$ (363)

Similarly,

$$\cos(B - E) = \frac{\sin^2 \tfrac{1}{2} a + \sin^2 \tfrac{1}{2} a - \sin^2 \tfrac{1}{2} b}{2 \sin \tfrac{1}{2} a \cos \tfrac{1}{2} b \sin \tfrac{1}{2} c},$$ (364)

and

$$\cos(C - E) = \frac{\sin^2 \tfrac{1}{2} a + \sin^2 \tfrac{1}{2} b - \sin^2 \tfrac{1}{2} c}{2 \sin \tfrac{1}{2} a \sin \tfrac{1}{2} b \cos \tfrac{1}{2} c}.$$ (365)

From the formulae (360)-(362) we get, by transformation, the following values for $\cot(A - E)$, viz.,

$$\cot(A - E) = \frac{\cot \tfrac{1}{2} a \tan \tfrac{1}{2} b}{\sin C} - \cot C$$ (366)

$$= \frac{\tan \tfrac{1}{2} b \tan \tfrac{1}{2} c}{\sin A} + \cot A$$ (367)

$$= \frac{\tan \tfrac{1}{2} a \cot \tfrac{1}{2} b}{\sin B} - \cot B,$$ (368)

with similar values for $\cot(B - E)$, $\cot(C - E)$.

84. *To find* $\sin \tfrac{1}{2} E$, $\cos \tfrac{1}{2} E$, $\tan \tfrac{1}{2} E$.

$$\sin \tfrac{1}{2} E = \sqrt{\frac{1 - \cos E}{2}}$$

$$= \sqrt{\frac{1 - \cos^2 \tfrac{1}{2} a - \cos^2 \tfrac{1}{2} b - \cos^2 \tfrac{1}{2} c + 2 \cos \tfrac{1}{2} a \cos \tfrac{1}{2} b \cos \tfrac{1}{2} c}{4 \cos \tfrac{1}{2} a \cos \tfrac{1}{2} b \cos \tfrac{1}{2} c}}$$

(from (357))

$$= \sqrt{\frac{\sin \tfrac{1}{2} s \sin \tfrac{1}{2}(s - a) \sin \tfrac{1}{2}(s - b) \sin \tfrac{1}{2}(s - c)}{\cos \tfrac{1}{2} a \cos \tfrac{1}{2} b \cos \tfrac{1}{2} c}}.$$ (369)

Similarly,

$$\cos \tfrac{1}{2} E = \sqrt{\frac{\cos \tfrac{1}{2} s \cos \tfrac{1}{2}(s - a) \cos \tfrac{1}{2}(s - b) \cos \tfrac{1}{2}(s - c)}{\cos \tfrac{1}{2} a \cos \tfrac{1}{2} b \cos \tfrac{1}{2} c}}.$$ (370)

Hence

$$\tan \tfrac{1}{2} E = \sqrt{\tan \tfrac{1}{2} s \tan \tfrac{1}{2}(s - a) \tan \tfrac{1}{2}(s - b) \tan \tfrac{1}{2}(s - c)}.$$ (371)

The same value as that obtained in § 48 by a different method.

85. If we put

$$L = \sqrt{\cot \tfrac{1}{2} s \, \tan \tfrac{1}{2} (s - a) \, \tan \tfrac{1}{2} (s - b) \, \tan \tfrac{1}{2} (s - c)}$$

(see equation (186)), we have

$$\tan \tfrac{1}{2} E = \frac{L}{\cot \tfrac{1}{2} s}. \tag{372}$$

Hence (§ 78),

$$\tan \tfrac{1}{2} (A - E) = \frac{L}{\tan \tfrac{1}{2} (s - a)}, \tag{373}$$

$$\tan \tfrac{1}{2} (B - E) = \frac{L}{\tan \tfrac{1}{2} (s - b)}, \tag{374}$$

$$\tan \tfrac{1}{2} (C - E) = \frac{L}{\tan \tfrac{1}{2} (s - c)}. \tag{375}$$

Compare equations (182)–(185.)

86. *Lhuilier's* theorem can be proved, independently of $\sin \tfrac{1}{2} E$, $\cos \tfrac{1}{2} E$, as follows. Thus :—

$$\tan \tfrac{1}{2} E = \frac{\sin \tfrac{1}{4}(A + B + C - \pi)}{\cos \tfrac{1}{4}(A + B + C - \pi)} = \frac{\sin \tfrac{1}{2}(A + B) - \sin \tfrac{1}{2}(\pi - C)}{\cos \tfrac{1}{2}(A + B) + \cos \tfrac{1}{2}(\pi - C)}$$

$$= \frac{\sin \tfrac{1}{2}(A + B) - \cos \tfrac{1}{2} C}{\cos \tfrac{1}{2}(A + B) + \sin \tfrac{1}{2} C} = \frac{\cos \tfrac{1}{2}(a - b) - \cos \tfrac{1}{2} c}{\cos \tfrac{1}{2}(a + b) + \cos \tfrac{1}{2} c} \cdot \frac{\cos \tfrac{1}{2} C}{\sin \tfrac{1}{2} C}$$

(by Delambre's Analogies)

$$= \frac{\sin \tfrac{1}{2}(s - a) \, \sin \tfrac{1}{2}(s - b)}{\cos \tfrac{1}{2} s \, \cos \tfrac{1}{2}(s - c)} \cot \tfrac{1}{2} C$$

$$= \sqrt{\tan \tfrac{1}{2} s \, \tan \tfrac{1}{2} (s - a) \, \tan \tfrac{1}{2} (s - b) \, \tan \tfrac{1}{2} (s - c)}.$$

87. Prouhet's proof of *Lhuilier's* theorem.

From the third of Delambre's Analogies we have

$$\frac{\sin \left(\dfrac{C}{2} - E \right)}{\sin \dfrac{C}{2}} = \frac{\cos \left(\dfrac{a + b}{2} \right)}{\cos \dfrac{c}{2}} ;$$

therefore

$$\frac{\sin \frac{C}{2} + \sin\left(\frac{C}{2} - E\right)}{\sin \frac{C}{2} - \sin\left(\frac{C}{2} - E\right)} = \frac{\cos \frac{c}{2} + \cos \frac{a+b}{2}}{\cos \frac{c}{2} - \cos \frac{a+b}{2}}.$$

Hence $\tan \frac{1}{2} E \cot \frac{1}{2} (C - E) = \tan \frac{s}{2} \tan \frac{1}{2} (s - c).$ (376)

Similarly, from the first of Delambre's Analogies,

$$\tan \tfrac{1}{2} E \tan \tfrac{1}{2} (C - E) = \tan \tfrac{1}{2} (s - a) \tan \tfrac{1}{2} (s - b). \quad (377)$$

Hence, multiplying and extracting square root, &c.   See *Nouvelles Annales*, 1856, p. 91.

Exercises XXV.

1. If the arc AD, drawn from A to a point D in the side BC, bisect the area of the spherical triangle, prove that

$$\cos \tfrac{1}{2} AB : \cos \tfrac{1}{2} AC : : \sin \tfrac{1}{2} BD : \sin \tfrac{1}{2} DC. \quad (378)$$

2. If ABC be a triangle, having $a = b = \dfrac{\pi}{3}$, $c = \dfrac{\pi}{2}$, prove $\sin E = \dfrac{1}{3}$.

3. If in fig., § 79, the arc EM be cut off equal to $\frac{1}{2} AB$, and MN be drawn perpendicular to the great circle DE; then $MN = E$. (Gudermann.)

4. Prove that $\sin (A - E) = \dfrac{\cos \frac{1}{2} b \cos \frac{1}{2} c \sin A}{\cos \frac{1}{2} a}$.   (379)

5. Prove that

$$\cos (A - E) = \frac{\sin \frac{1}{2} b \sin \frac{1}{2} c + \cos \frac{1}{2} b \cos \frac{1}{2} c \cos A}{\cos \frac{1}{2} a}. \quad (380)$$

6. Prove that in a right-angled spherical triangle $\tan E = \tan \frac{1}{2} a \tan \frac{1}{2} b$.

7. If a', b', c', A', B', C' denote the sides and angles of the triangle supplemental to ABC, prove

$$\cot \tfrac{1}{2} s \cot \tfrac{1}{2} s' = \tan \tfrac{1}{2} (s - a) \tan \tfrac{1}{2} (s' - a'). \quad (381)$$

8. In the same case, if $2E'$ denote the spherical excess of the polar triangle, prove that

$$\tan \tfrac{1}{2} E \tan \tfrac{1}{2} E'' = \tan \tfrac{1}{2} (A - E) \tan \tfrac{1}{2} (A' - E'). \quad (382)$$

9. Prove that the arc joining the middle points of any two sides of a spherical triangle is less than a quadrant.

10. The cosines of the arcs joining the middle points of the sides of a spherical triangle are proportional to the cosines of half the sides.

11. Solve a spherical triangle, being given a, $b \pm c$, and E.

12. If s denote the semiperimeter of a spherical triangle, Δ, Δ_a, Δ_b, Δ_c its area, and the areas of its colunar triangles; prove that

$$\tan^2 \frac{s}{2} = \tan \tfrac{1}{4} \Delta \cdot \cot \tfrac{1}{4} \Delta_a \cdot \cot \tfrac{1}{4} \Delta_b \cdot \cot \tfrac{1}{4} \Delta_c. \tag{383}$$

13. Prove $\sin E = \cot R \tan \tfrac{1}{2} a \, \tan \tfrac{1}{2} b \, \tan \tfrac{1}{2} c$.   (384)

14. ,, $\sin s = \sin a \, \cos \tfrac{1}{2} B \, \cos \tfrac{1}{2} C \div \sin \tfrac{1}{2} A$.   (385)

15. If $(a + b + c) = \pi$, prove $\cos A + \cos B + \cos C = 1$, $\cos A$.   (385)

16. In the same case, prove that $\cos A = \tan \dfrac{B}{2} \, \tan \dfrac{C}{2}$.   (386)

17. Prove $\cos s = \dfrac{\sin^2 \tfrac{1}{2} A + \sin^2 \tfrac{1}{2} B + \sin^2 \tfrac{1}{2} C - 1}{2 \sin \tfrac{1}{2} A \, \sin \tfrac{1}{2} B \, \sin \tfrac{1}{2} C}$.   (387)

18. ,, $\sin^2 \dfrac{s}{2} = \dfrac{\sin \tfrac{1}{2} E \, \cos \tfrac{1}{2}(A - E) \, \cos \tfrac{1}{2}(B - E) \, \cos \tfrac{1}{2}(C - E)}{4 \sin \tfrac{1}{2} A \, \sin \tfrac{1}{2} B \, \sin \tfrac{1}{2} C}$.   (388)

19. If $E = \tfrac{1}{2} \pi$, prove that $\cos A = - \cot \dfrac{b}{2} \, \cot \dfrac{c}{2}$.   (389)

20. Prove $\cos (s - a) = \dfrac{\cos^2 \tfrac{1}{2} B + \cos^2 \tfrac{1}{2} C - \cos^2 \tfrac{1}{2} A}{2 \sin \tfrac{1}{2} A \, \cos \tfrac{1}{2} B \, \cos \tfrac{1}{2} C}$.   (390)

21. If I be incentre of a spherical triangle, prove that

$$\cos BIC = \dfrac{\cos^2 \dfrac{A}{2} - \cos^2 \dfrac{B}{2} - \cos^2 \dfrac{C}{2}}{2 \cos \dfrac{B}{2} \, \cos \dfrac{C}{2}}. \qquad \text{(Neuberg.)} \tag{391}$$

22. If I_a, I_b, I_c be the incentres of the triangles colunar to ABC, prove that

$$\cos BI_a C = \dfrac{\cos^2 \dfrac{A}{2} - \sin^2 \dfrac{B}{2} - \sin^2 \dfrac{C}{2}}{2 \sin \dfrac{B}{2} \, \sin \dfrac{C}{2}}. \qquad \text{(\textit{Ibid.})} \tag{392}$$

23. The angle $BI_a C$ corresponds in the supplemental triangle to the arc joining the middle points of two sides.   (*Ibid.*)

24. If I_a be the incentre of the colunar triangle $A'BC$, from I_a let fall perpendiculars $I_a D$, $I_a E$, $I_a F$ on the sides BC, CA, AB, respectively; then the angle $BI_a C = \tfrac{1}{2} FI_a E = FI_a A$. The triangle $FI_a A$ gives

$$\cos FI_a A = \cos AF \cdot \sin FAI_a = \cos s \cdot \sin \tfrac{1}{2} A.$$

Hence $\cos BI_a C = \cos s \cdot \sin \tfrac{1}{2} A.$

Hence, from (392), we get

$$\cos s = \frac{\cos^2 \dfrac{A}{2} - \sin^2 \dfrac{B}{2} - \sin^2 \dfrac{C}{2}}{2 \sin \frac{1}{2} A \, \sin \frac{1}{2} B \, \sin \frac{1}{2} C} \cdot \qquad \text{(Neuberg.)} \quad (393)$$

This theorem is the correlative of 357. We can get, in the same manner,

$$\cos (s - a), \;\; \cos (s - b), \;\; \cos (s - c) \, ; \quad \cos \frac{s}{2}, \;\; \sin \frac{s}{2}, \;\; \tan \frac{s}{2}, \;\; \tan \frac{s - a}{2}, \;\; \&c.$$

SECTION II.—LEXELL'S THEOREM.

88. *If the base BC of a spherical triangle ABC be given in magnitude and position, and the spherical excess in magnitude, the locus of the vertex is a small circle of the sphere.*

STEINER'S PROOF.   *Lemma.*—If upon the base BC a spherical triangle be constructed, such that $A - E$ is given, the locus of A is a small circle, namely, the circumcircle of the triangle. For if O be the pole of the circumcircle (see fig., § 75), the angle $OBC = OCB = (A - E)$. Hence O is a given point, and the circle is given in position.

LEXELL'S THEOREM.—Let ABC be one position of the triangle, $2E$ the spherical excess constant. Let the points B', C' be the

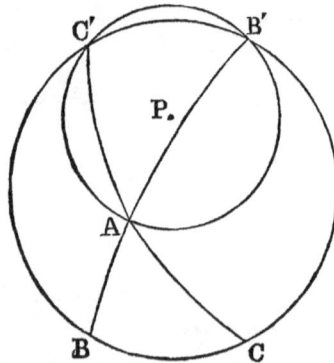

Fig. 32.

antipodes of B, C; let P be the pole of the circle $AB'C'$; then we have $2E = A + B + C - \pi = B'AC' + \pi - AB'C' + \pi$

$- A\,C'B' - \pi = \pi + A - B' - C'$.   Hence $B' + C' - A$ is constant; and by the lemma the locus of A is the circumcircle of the triangle $B'\,C'A$.

Or thus : SERRET's PROOF.—Let, as before, B', C' be the antipodes of B, C; let E' be the spherical excess of $B'C'A$, and R' its circumradius ; then we have (§ 75),

$$\tan R' = \tan \tfrac{1}{2} a \div \sin (A - E') = \tan \tfrac{1}{2} a \div \sin E.$$

Hence since a and E are given in magnitude, R' is given in magnitude, and the circumcircle of $B'C'A$ is evidently given in position, and is the locus required.

89. Steiner's Theorem.—*The great circles through angular points of a spherical triangle ABC, and which bisect its area, are concurrent.*   Let the circles bisecting the area meet the opposite sides in the points a, β, γ, respectively ; also, let A', B', C' be the antipodes of A, B, C.   Now the areas of the triangles ABa, $AB\beta$ are equal, each being half of ABC.   Hence, by Lexell's theorem, the points A', B', a, β are concyclic.   Similarly, each of the systems of points B', C', β, γ ; C', A', γ, a, is concyclic.

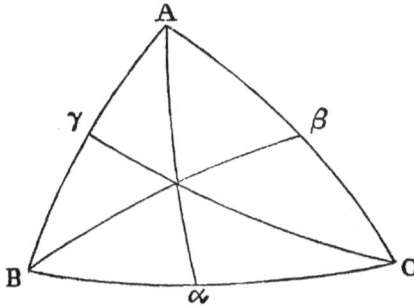

Fig. 33.

Let the point common to the planes of these three small circles be P, then the lines of intersection of these planes two by two pass through P.   Hence, if O be the centre of the sphere, the planes $OB'\beta B$, $OC''\gamma C$, $OA'aA$ have a common line of intersection, namely, the line OP.   Hence the proposition is proved.

Analytical Proof.—The triangles ABa, ACa having the same spherical excess, we have, by Cagnoli's theorem, § 80,

$$\frac{\sin \frac{c}{2} \sin \frac{Ba}{2} \sin B}{\cos \frac{1}{2} Aa} = \frac{\sin \frac{b}{2} \sin \frac{Ca}{2} \sin C}{\cos \frac{1}{2} Aa} .$$

Hence

$$\frac{\sin \frac{Ba}{2}}{\sin \frac{Ca}{2}} = \frac{\cos \frac{c}{2}}{\cos \frac{b}{2}} ;$$

and from this and two similar equations we get

$$\sin \tfrac{1}{2} Ba \, \sin \tfrac{1}{2} C\beta \, \sin \tfrac{1}{2} A\gamma = \sin \tfrac{1}{2} aC \, \sin \tfrac{1}{2} \beta A \, \sin \tfrac{1}{2} \gamma B.$$

$$(1)$$

Also the triangles ABa, γBC having equal areas,

$$\tan \tfrac{1}{2} c \, \tan \tfrac{1}{2} Ba = \tan \tfrac{1}{2} \gamma B \, \tan \tfrac{1}{2} a. \quad \text{(Art. 81, } \textit{Cor.} \text{ 2.)}$$

Hence,

$$\tan \tfrac{1}{2} Ba \, \tan \tfrac{1}{2} C\beta \, \tan \tfrac{1}{2} A\gamma = \tan \tfrac{1}{2} aC \, \tan \tfrac{1}{2} \beta A \, \tan \tfrac{1}{2} \gamma B.$$

$$(2)$$

From (1) and (2) we have

$$\cos \tfrac{1}{2} Ba \, \cos \tfrac{1}{2} C\beta \, \cos \tfrac{1}{2} A\gamma = \cos \tfrac{1}{2} aC \, \cos \tfrac{1}{2} \beta A \, \cos \tfrac{1}{2} \gamma B. \quad (3)$$

From (1) and (3) we get

$$\sin Ba \, \sin C\beta \, \sin A\gamma = \sin aC \, \sin \beta A \, \sin \gamma B. \quad (4)$$

Hence the arcs Aa, $B\beta$, $C\gamma$ are concurrent. (NEUBERG.)

Cor.—The triangular co-ordinates of the point of intersection of the arcs Aa, $B\beta$, $C\gamma$ are

$$\cos \tfrac{1}{2} a - \cos \tfrac{1}{2} b \, \cos \tfrac{1}{2} c, \quad \cos \tfrac{1}{2} b - \cos \tfrac{1}{2} c \, \cos \tfrac{1}{2} a,$$

$$\cos \tfrac{1}{2} c - \cos \tfrac{1}{2} a \, \cos \tfrac{1}{2} b. \quad \textit{(Ibid.)}$$

These values are obtained from the equation

$$\frac{\sin \tfrac{1}{2} Ba}{\sin \left(\tfrac{1}{2} a - \tfrac{1}{2} Ba \right)} = \frac{\cos \tfrac{1}{2} c}{\cos \tfrac{1}{2} b} ,$$

which gives $\cot \tfrac{1}{2} Ba$, and thus $\sin Ba$.

90. Keogh's Theorem.—*The sine of half the spherical excess is equal to twice the Staudtian of the triangle formed by joining the middle points of the sides.*—*Nouv. Annales*, 1857, p. 142.

Dem.—Let A', B', C' be the middle points of the sides. (See figure, § 79.) Then we have, from the right-angled triangle DAF,

$$\cos DAF = \sin D \cos DF, \text{ equation (111)};$$

that is, $\qquad\qquad \sin E = \sin D \sin B'A'.$

But $\sin D$ = sine of the perpendicular from C' on $B'A'$. Hence, if n' denote the Staudtian of $A'B'C'$, we have

$$\sin E = 2n'. \tag{364}$$

91. *To find the triangle of maximum area, two sides, b, c, being given.*

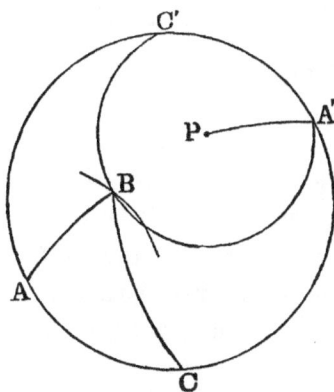

Fig. 34.

Sol.—The following is Steiner's geometrical solution :—

Suppose the side AC to be fixed in position. Let $A'C'$ be the antipodes of A and C; through A', C' let a small circle be described with pole P, such that the angle

$$PA'C' = \frac{\pi}{2} - E;$$

then every triangle having AC as base, and vertex any point on the small circle, will have a constant area, namely, $2Er^2$. If the side AB be given, the small circle BB_1 described with A as pole, and spherical radius c, cuts Lexell's circle in the points B, B_1, each of the two triangles ABC, AB_1C will have an area $= 2Er^2$.

In order that the problem may be possible, Lexell's circle must meet the circle BB_1; or, what is the same thing, the angle $PA'C'$ equal to $\dfrac{\pi}{2} - E$, must be sufficiently large, the minimum of $\dfrac{\pi}{2} - E$, or the maximum of E, corresponding to the case where the small circles touch each other. Then the points A, B, P, A' are on the same great circle, and the triangle $BC'A'$ is a diametral triangle;

$$\therefore \quad C' = A' + B;$$

but $\qquad A' = \pi - A, \quad \text{and} \quad C' = \pi - C.$

Hence $A = B + C$, and the required triangle is diametral, a being the diameter.

Cor.—If AB be greater than AC', the circle BB_1 must intersect Lexell's circle, and there will be neither a maximum nor a minimum; but if AB be greater than AC', $AB + AC$ will be greater than AA', or $b + c$ greater than π. Hence, if $b + c > \pi$, there will be neither a maximum nor a minimum.

Trigonometrical Solution (Neuberg's).

$1°$. We have, by Cagnoli's formulae (351), (352),

$$\sin (A - E) = \sin E \cot \tfrac{1}{2}b \, \cot \tfrac{1}{2}c.$$

If $\qquad \cot \tfrac{1}{2}b \cot \tfrac{1}{2}c > 1, \quad \text{or} \quad b + c > 180,$

$\sin E$ may have any value, and then $\sin (A - E)$ may be found, and the triangle is possible. Hence there is neither a maximum nor a minimum.

2°. If $\qquad \cot \tfrac{1}{2} b \cdot \cot \tfrac{1}{2} c = 1$, or $b + c = \pi$,

we have $A - E = E$, and the triangle becomes a lune formed by the circle $C'AC$, and the greater circle tangential to Lexell's circle in C'.

3°. If $\cot \tfrac{1}{2} b \cdot \cot \tfrac{1}{2} c < 1$, or $b + c < \pi$, $\sin E$ is no longer arbitrary. In order that $\sin (A - E) < 1$, $\sin E$ must be $< \tan \tfrac{1}{2} b \tan \tfrac{1}{2} c$. The maximum of E corresponds to $\sin E = \tan \tfrac{1}{2} b \cdot \tan \tfrac{1}{2} c$, and then

$$A - E = \frac{\pi}{2}, \quad \text{or} \quad A = B + C.$$

Exercises.—XXVI.

1. Construct a lune equal in area to a given triangle (make use of Lexell's circle).

2. Construct by means of Lexell's theorem a triangle AB_1C_1 equal in area to a given triangle ABC, and having two given sides b_1, c_1.

3. Construct on the side BC of a given triangle ABC an equivalent isosceles triangle.

4. Convert a triangle ABC into an equivalent isosceles triangle, having a common angle A.

5. Transform a spherical polygon $ABCDE$ into an equivalent spherical triangle.

[Employ Lexell's circle in the same manner as parallel lines are employed in the corresponding question in *Plane Geometry*.]

6. Being given a spherical polygon $ABCDE$, if the sides be produced in the same sense, and with each vertex as pole, an arc of a great circle be described, limited by the sides of the corresponding exterior angle of the polygon, prove that the total figure thus formed is equal in area to a hemisphere.—(Neuberg.)

7. Being given A and E, prove that, if a is a minimum, $b = c$.

We have $\qquad \sin E = \dfrac{\sin \tfrac{1}{2} b \sin \tfrac{1}{2} c \sin A}{\cos \tfrac{1}{2} a}.$

Hence, from the required condition, $\sin \tfrac{1}{2} b \sin \tfrac{1}{2} c$ is a maximum; but

H

$\tan \frac{1}{2} b \tan \frac{1}{2} c$ is constant (§ 82, *Cor.* 2) ; therefore $\cos \frac{1}{2} b \cos \frac{1}{2} c$ is a maximum ;

$$\therefore \quad \sec \tfrac{1}{2} b \sec \tfrac{1}{2} c = \sqrt{\ \overline{(\tan \tfrac{1}{2} b - \tan \tfrac{1}{2} c)^2 + (1 + \tan \tfrac{1}{2} b \tan \tfrac{1}{2} c)^2}}$$

is a minimum. Hence $b = c$.—(NEUBERG.)

8. Being given A and E, prove that if $b + c$ be a maximum, $b = c$,

$$\tan \tfrac{1}{2} (b + c) = \frac{\tan \tfrac{1}{2} b + \tan \tfrac{1}{2} c}{1 - \tan \tfrac{1}{2} b \tan \tfrac{1}{2} c}. \quad \text{Hence } b = c.\text{—}(\textit{Ibid.})$$

9. If ABC, ABD be two spherical triangles of equal areas on the same side of the common base AB, prove that

$$\sin \tfrac{1}{2} AB . \sin \tfrac{1}{2} CD + \cos \tfrac{1}{2} AC . \cos \tfrac{1}{2} BD = \cos \tfrac{1}{2} AD . \cos \tfrac{1}{2} BC.$$

10. Investigate the maximum or minimum of E, being given A and $b + c$.

$$\cot E = \frac{\cot \tfrac{1}{2} b . \cot \tfrac{1}{2} c}{\sin A} + \cot A = \left\{ 1 + \frac{2 \cos \tfrac{1}{2} (b + c)}{\cos \tfrac{1}{2} (b - c) - \cos \tfrac{1}{2} (b + c)} \right\} \frac{1}{\sin A} + \cot A$$

$$= \frac{2 \cos \tfrac{1}{2} (b + c)}{\sin A \left\{ \cos \tfrac{1}{2} (b - c) - \cos \tfrac{1}{2} (b + c) \right\}} + \cot \frac{A}{2}.$$

If $\tfrac{1}{2} (b + c) < 90°$, then $\cos \tfrac{1}{2} (b - c) - \cos \tfrac{1}{2} (b + c) > 0$, and the minimum of $\cot E$ or the maximum of E corresponds to $\cos \tfrac{1}{2} (b - c) = 1$ or to $b = c$. If $\tfrac{1}{2} (b + c) > 90°$, in the colunar triangle $A'BC$, $\tfrac{1}{2} (b' + c') < 90°$; and since A and $b + c$ are constant in ABC, A and $b' + c'$ are constant in BCA'. Hence the area of BCA' is a maximum when $b' = c'$; and therefore when $\tfrac{1}{2} (b + c) > 90°$, the area of ABC is a minimum when $b = c$.—(*Ibid.*)

11. If $E = \dfrac{\pi}{2}$, prove that $\cos A = - \cot \tfrac{1}{2} b \cot \tfrac{1}{2} c$.

12. If O be a point such that the areas of the triangles AOB, BOC, COA are equal, prove that

$$\tan \frac{OA}{2} : \tan \frac{OB}{2} : \tan \frac{OC}{2} :: \sin \left(BOC - \frac{E}{3} \right) : \sin \left(COA - \frac{E}{3} \right)$$

$$: \sin \left(AOB - \frac{E}{3} \right).$$

13. In the same case prove that the small circle passing through the antipodes of O, and the extremities of any side of the spherical triangle, intersects that side at an angle $= E \div 3$.

CHAPTER VI.

SMALL CIRCLES ON THE SPHERE.

SECTION I.—COAXAL CIRCLES.

92. *If an arc of a great circle passing through a fixed point O cut a small circle X in the points A, B, $\tan \frac{1}{2} AO \cdot \tan \frac{1}{2} OB$ is constant.*

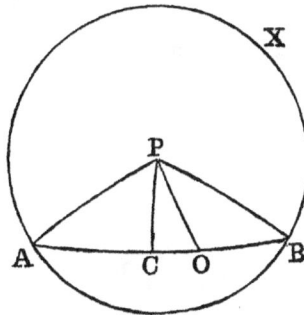

Fig. 35.

DEM.—Let P be the pole of the small circle. Join OP. Let fall the perpendicular PC; then, from the triangles ACP, OCP, we have

$$\frac{\cos AC}{\cos CO} = \frac{\cos AP}{\cos OP}.$$

Hence
$$\tan \tfrac{1}{2}(AC + CO)\tan \tfrac{1}{2}(AC - CO)$$
$$= \tan \tfrac{1}{2}(AP + PO)\tan \tfrac{1}{2}(AP - PO);$$

or, denoting the radius of X by ρ and OP by δ,

$$\tan \tfrac{1}{2} OA \cdot \tan \tfrac{1}{2} OB = \tan \tfrac{1}{2}(\rho + \delta)\tan \tfrac{1}{2}(\rho - \delta). \quad (395)$$

H 2

Def. XXVII.—*The product* $\tan \frac{1}{2} OA \cdot \tan \frac{1}{2} OB$ *is called the spherical power of O, with respect to the circle. It is positive or negative, according as O is exterior or interior to the circle.*

Cor. 1.—If δ be the distance of a point O from the pole of a small circle, radius ρ, the spherical power

$$= \frac{\cos \rho - \cos \delta}{\cos \rho + \cos \delta}. \tag{396}$$

Cor. 2.—If from any point O outside a small circle two arcs be drawn to it, of which one, OD, is a tangent, and the other a secant, meeting it in the points A, B; then

$$\tan^2 \tfrac{1}{2} OD = \tan \tfrac{1}{2} OA \cdot \tan \tfrac{1}{2} OB.$$

93. *If two small circles cut orthogonally, the plane of either passes through the vertex of the cone, touching the sphere along the other.*

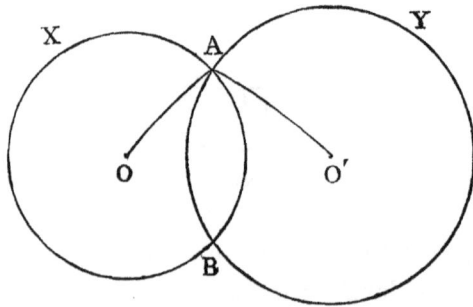

Fig. 36.

Dem.—Let the circles be X, Y; O, O' their spherical centres, A, B their points of intersection; then it is evident that the tangent line to the arc AO is in the plane of Y, and that it passes through the vertex of the cone, which touches the sphere along the circumference of X. Hence the proposition is proved.

Cor.—If any number of circles on the sphere have a common orthogonal circle, their planes pass through a common point;

and conversely, if the planes of any number of circles pass through a common point, they have a common orthogonal circle.

94. *If the planes of a system of circles S have a common line L of intersection, the circles S have an infinite number of common orthogonal circles.*

Dem.—Take any point P in the line L, and through it draw tangent lines to the sphere; these will touch along a circle which (§ 93) cuts each circle of the system S orthogonally; and since the same thing holds for each point on L, we have an infinite number of circles forming a system S', each of which cuts each circle of S orthogonally.

Cor.—The planes of the circles of the system S' have a common line of intersection.

For, take any two of them, say P and Q. Now (§ 93) the plane of each passes through the vertex of each of the cones, touching the sphere along the circles of the system S. Hence the vertices are collinear, and the plane of each circle of S' passes through the line of collinearity.

Def. XXVIII.—*A system of circles S, whose planes pass through a common line L, is called a* coaxal system.

Def. XXIX.—*The circle of the system S, whose plane passes through the centre of the sphere, is called the* radical circle *of the system.*

Def. XXX.—*If through L two tangent planes be drawn to the sphere, their points of contact, regarded as infinitely small circles, are the limiting points of the system.*

Cor.—Each circle of the system S' passes through the limiting points of S.

95. *If X, Y, Z be three circles of a coaxal system, and from any point P in X tangents PT, PT' be drawn to Y and Z; then $\sin \frac{1}{2} PT : \sin \frac{1}{2} PT'$ in a given ratio.*

DEM.—Let O, O' be the spherical centres of Y, Z. Join OP, $O'P$ by arcs of great circles; then, if the radius of the sphere be unity, the perpendicular from P on the plane of $Y = \cos OT$ $- \cos OP = \cos OT - \cos OT \cdot \cos PT = \cos OT \cdot 2 \sin^2 \frac{1}{2} PT$. Similarly, the perpendicular from P on the plane of $Z = \cos O'T' \cdot 2 \sin^2 \frac{1}{2} PT'$. But since the planes of X, Y, Z are collinear, the perpendiculars have a given ratio. Hence the ratio of $\cos OT \cdot \sin^2 \frac{1}{2} PT : \cos O'T' \cdot \sin^2 \frac{1}{2} PT'$ is given, and OT, $O'T'$ are given, being the spherical radii of Y and Z. Hence the ratio of $\sin \frac{1}{2} PT : \sin \frac{1}{2} PT'$ is given.

Cor.—If $PT = PT'$, the locus of P is the radical circle of the system.

EXERCISES.—XXVII.

1. The radical circles of three small circles taken in pairs are concurrent.

2. If there be a coaxal system of circles S, and a circle X distinct from it, then the radical circles of X, combined with each circle of S, are concurrent.

3. If through a point on the radical circle of two small circles we draw a spherical secant to each, the four points of intersection are concyclic.

4. Through two points of the sphere describe a small circle touching a given great circle.

5. If through a fixed point A we draw a great circle, cutting a given small circle in the points B, C, and if a point D be taken on it, such that $\tan^2 \frac{1}{2} AD = \tan^2 \frac{1}{2} DB \cdot \tan^2 \frac{1}{2} DC$, prove that the locus of D is a great circle.

6. If X, Y be two small circles; PT, PT' two tangents to them from a point P, prove that the locus of P is a circle, if $m \cos PT + n \cos PT'$ be constant, m and n being given numbers.

7. The locus of the poles of small circles, intersecting two small circles X, Y at the extremities of two spherical diameters, is the radical circle of X, Y.

8. Describe a circle cutting three small circles at the extremities of three spherical diameters.

9. If two rectangular secants intersecting in M cut a small circle in the pairs of points A, B; C, D, prove that

$$\tan^2 \tfrac{1}{2} MA + \tan^2 \tfrac{1}{2} MB + \tan^2 \tfrac{1}{2} MC + \tan^2 \tfrac{1}{2} MD = \frac{4 \sin^2 \rho}{(\cos \rho + \cos \delta)^2}, \quad (397)$$

where δ is the distance of M from the pole of the small circle.—(NEUBERG.)

10. If from any point P of a great circle MP tangents PT, PT' be drawn to a small circle, prove that $\tan \tfrac{1}{2} MPT \cdot \tan \tfrac{1}{2} MPT''$ is constant.

11. The difference of the cosines of the tangent arcs, drawn from any point P on the surface of a sphere to two small circles X, Y, is proportional to the sine of the perpendicular drawn from P to the radical axis of X and Y.

SECTION II.—CENTRES OF SIMILITUDE.

96. DEF. XXXI.—*Two points, S, S', which divide the arc PP', joining the poles of two small circles Y, Z externally and internally in the spherical ratio of the sines of the radii, are called the centres of similitude of the small circles.*

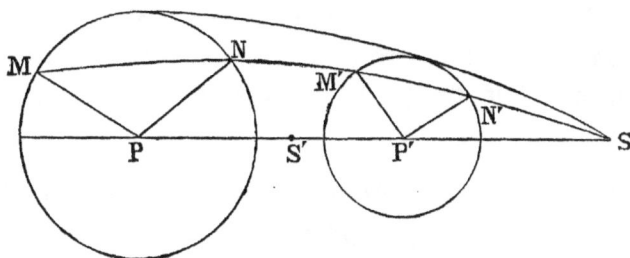

Fig. 37.

Cor.—Common tangents to the small circles pass through the centres of similitude, viz., the direct common tangents through the external centre, and the inverse common tangent through the internal centre.

DEF. XXXII.—*If through a centre of similitude we draw a secant cutting the circles, then the pairs of points M, M'; N, N' are said to be homothetic, and M, N'; M', N are inverse.*

97. *If the secant through a centre of similitude S meets the circles in the homothetic points M, M'; then tan ½ SM : tan ½ SM' in a given ratio.*

DEM.—From the definition

$$\sin SP : \sin PM :: \sin SP' : \sin P'M.$$

Hence it follows that the angle $SMP = SM'P'$.

Now since the triangles SMP, $SM'P'$ have two angles in one respectively equal to two angles in the other, it follows from the third of Napier's Analogies that

$$\tan \tfrac{1}{2} SM : \tan \tfrac{1}{2} SM' :: \tan \tfrac{1}{2}(SP + PM) : \tan \tfrac{1}{2}(SP' + P'M');$$

that is, in a given ratio.   Similarly,

$$\tan \tfrac{1}{2} SN : \tan \tfrac{1}{2} SN' \text{ in a given ratio.} \qquad (398)$$

Cor.—

$\tan \tfrac{1}{2} SM . \tan \tfrac{1}{2} SN'$ is constant, as also $\tan \tfrac{1}{2} SN . \tan \tfrac{1}{2} SM'$.

This follows from §§ 92, · 97. (Compare *Sequel*, Prop. II., page 83.

Cor.—If there be given a point S and a circle Y, and on the arc SM joining S to any point M on Y a point N' be taken, such that $\tan \tfrac{1}{2} SM . \tan \tfrac{1}{2} SN'$ is constant, the locus of N' is a circle.

98. *The six centres of similitude of three small circles taken in pairs lie three by three on four great circles, called axes of similitude of the small circles.*

DEM.—If a, b, c be the radii of the circles; A, B, C their spherical centres, $A'B'C'$ the internal centres of similitude, and $A''B''C''$ the externals; then we have by definitions

$$(AB, C'') = a \div b, \quad (BC, A'') = b \div c, \quad (CA, B'') = c \div a.$$

Hence $(AB, C'') . (BC, A'') . (CA, B'') = 1.$

Hence (§ 70) the points A'', B'', C'' lie on a great circle. Similarly, it may be shown that any two internal centres and an external centre lie on a great circle.

Cor. 1—If a variable circle touch two fixed circles, the great circle passing through the points of contact passes through a fixed point, namely, a centre of similitude of the two circles ; for the points of contact are centres of similitude.

Cor. 2.—If a variable circle touch two fixed circles, the tangent drawn to it from the centre of similitude, through which the chord of contact passes, is constant.

99. DEF. XXXIII.—*Being given a fixed point S, and any line whatever, γ, on the sphere, if upon the arc of a great circle joining S to any point M of γ a point M' be taken, such that $\tan \frac{1}{2} SM$: $\tan \frac{1}{2} SM'$ in a given ratio, the locus of M' is said to be homothetic to γ.*

DEF. XXXIV.—*If M' be taken, such that $\tan \frac{1}{2} SM \cdot \tan \frac{1}{2} SM'$ is constant, the locus of M' is called the inverse of γ.*

This method of inversion was first employed in the Author's *Memoir on Cyclides and Sphero-quartics.* (Read before the Royal Society in 1871.)

EXERCISES.—XXVIII.

1. If two small circles touch two others, the radical axis of either pair passes through a centre of similitude of the other.

2. The figure homothetic to a circle is a circle.

3. The inverse of a circle is a circle.

4-5. *S* being the centre of similitude of two circles ; *M*, *N* two inverse points on these circles—1°, the tangents at *M* and *N* intersect on the radical axis ; 2°, these points are points of contact of two circles touching the two given circles.

6. The angle of intersection of two circles on the sphere is equal to the angle of intersection of the circles inverse to them.

7. Any two circles can be inverted into two equal circles.

8. Any three circles can be inverted into three equal circles.

9. If two circles be the inverses of two others, then any circle touching three of them will also touch the fourth.

10. If two points be the inverses of two other points, the four points are concyclic.

11. If through the centre of similitude S (see fig., § 96) another great circle be drawn, intersecting the circles Y, Z in the points μ, μ'; ν', ν, corresponding to the points M, M', N', N, the systems of points M, N', μ, ν'; M', N, μ', ν; M, N', μ', ν; M', N, μ, ν', are each concyclic, and the planes of the four circles pass through a common point.

12. If a variable circle on the sphere touch two fixed circles, the sine of its radius has a constant ratio to the sine of the perpendicular drawn from its spherical centre to the radical axis of the fixed circles.

13. If a variable circle touch two fixed circles, the ratio of the sines of half the tangents drawn to it from the limiting points is constant.

14. If a variable circle touch two fixed circles of a coaxal system, it cuts any circle of the system at a constant angle.

15. The inverse of a coaxal system is a coaxal system.

16. The inverse of a system of great circles passing through two common points is a coaxal system.

17. The inverse of a system of small circles having a common spherical centre is a coaxal system.

Section III.—Poles and Polars.

100. Lemmas.—*If the segment AB be harmonically divided in the points C, D; and E the middle point of AB; then*

$$1°. \qquad tan^2 EB = tan\, EC \,.\, tan\, ED. \qquad (399)$$

$$2°. \qquad cot\, AB = \tfrac{1}{2}\left(cot\, AC + cot\, AD\right). \qquad (400)$$

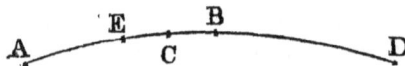

Fig. 38.

For, by definition,

$$\frac{\sin CA}{\sin CB} = \frac{\sin DA}{\sin DB};$$

$$\therefore \quad \frac{\sin CA - \sin CB}{\sin CA + \sin CB} = \frac{\sin DA - \sin DB}{\sin DA + \sin DB}.$$

Hence
$$\frac{\tan EC}{\tan EB} = \frac{\tan EB}{\tan ED}.$$

Hence the proposition is proved.

To prove 2°—We have
$$\sin BC \cdot \sin AD = \sin CA \cdot \sin DB,$$

or $\sin (AB - AC) \sin AD = \sin AC \sin (AD - AB).$

Hence, expanding and dividing by $\sin AB \sin AC \sin AD$, we get
$$\cot AC - \cot AB = \cot AB - \cot AD,$$

or $\cot AB = \tfrac{1}{2} (\cot AC + \cot AD).$

DEF. XXXV.—*Being given a small circle X, spherical centre P: if C, D be two points dividing the spherical diameter AB harmonically, an arc DD' of a great circle through one of these points, D, perpendicular to the diameter AB, is called the harmonic polar of the other, C, and C is called the harmonic pole of the arc DD'.*

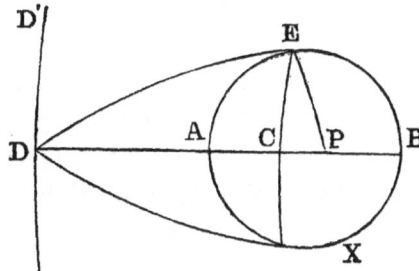

Fig. 39.

101. The arc of contact of spherical tangents, drawn from an exterior point D to a small circle X, is the harmonic polar of D.

DEM.—The right-angled triangles DEP, ECP give
$$\cos DPE = \frac{\tan EP}{\tan DP} = \frac{\tan CP}{\tan EP}.$$

Hence $\tan^2 EP$ or $\tan^2 EA = \tan DP \cdot \tan CP$; therefore the
points C, D are harmonic conjugates to A and B; and therefore, &c.

102. *If a spherical chord AB of a small circle X pass through a
fixed point C, the locus of the intersection of tangents AD, BD is
the harmonic polar of C.*

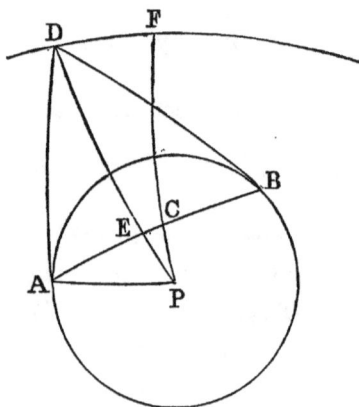

Fig. 40.

DEM.—Let P be the spherical centre of X. Join PD, PA, PC
by arcs of great circles, and let fall ·the perpendicular DF
on PC, produced if necessary. Now because in the spherical
quadrilateral $DECF$ the angles E, F are right, we have, equation (134),

$$\tan PC \cdot \tan PF = \tan PE \cdot \tan PD .= \tan^2 PA.$$

Hence the proposition is proved.

Cor. 1.—If a variable point move along an arc of ·a great
circle, its harmonic polar passes through a given point.

Cor. 2.—If C be a fixed point in a small circle X; AC, CB
any two arcs of great circles at right angles to each other; the
chord AB passes through a fixed point.

DEM.—Let Y be the circumcircle of the colunar triangle $AC'B$, and OO' the spherical centres of X and Y; and since the angle C is right, the circles X, Y cut orthogonally; therefore AO', $O'B$ are tangents to X. Hence AB is the polar of

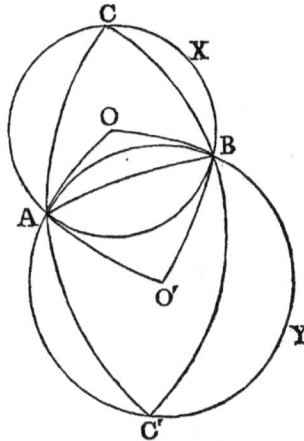

Fig. 41.

O', with respect to X; and since $O'A = O'C'$, the locus of O' is the radical axis of the circle X and the fixed point C'; and therefore AB, the polar of O' with respect to X, passes through a given point.

103. *Every secant (OA) passing through a given point O is cut harmonically by the circle X and the harmonic polar of O.*

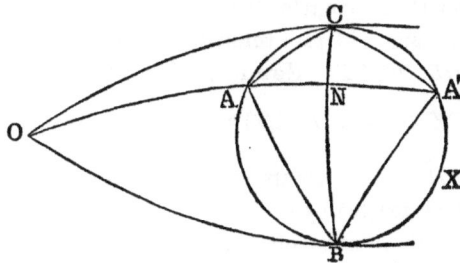

Fig. 42.

DEM.—Let BC be the polar of O, and let the sines of the

perpendiculars from A on the sides CO, OB, BC of the triangle OBC be denoted by x, y, z; and the sines of the perpendiculars from A' by x', y', z', respectively, then we have

$$x : z :: \sin OCA : \sin ACB;$$

that is, $$x : z :: \sin (B - E) : \sin C.$$

Similarly, $$y : z :: \sin (C - E) : \sin B;$$

$$\therefore \quad \frac{xy}{z^2} = \frac{\sin (B - E) \sin (C - E)}{\sin B \sin C} = \cos^2 \tfrac{1}{2} a.$$

In like manner, $$\frac{x'y'}{z'^2} = \cos^2 \tfrac{1}{2} a.$$

Hence $$xy : x'y' :: z^2 : z'^2;$$

but

$$xy : x'y' :: \sin^2 OA : \sin^2 OA', \text{ and } z^2 : z'^2 :: \sin^2 AN : \sin^2 A'N;$$

$$\therefore \quad \sin OA : OA' :: \sin AN : \sin NA'. \qquad (401)$$

Exercises.—XXIX.

1. If four points A, B, C, D lie on a great circle a, their anharmonic ratio is equal to that of their harmonic polars, with respect to any small circle X.

For if O be the spherical centre of X, P the harmonic pole of a, the perpendiculars from P on the circles OA, OB, OC, OD will be the harmonic polars of A, B, C, D, and will pass through the poles A', B', C', D' of the great circles OA, OB, OC, OD. Now it is evident that

$$(P - A'B'C'D') = (A'B'C'D') = (O - ABCD) = (ABCD).$$

2. If A, B, C, D be four points on a small circle X, and if the arcs AB, BC, CD, DA be denoted by a, b, c, d, respectively; then if P be any variable point on X, the anharmonic ratio

$$(P - ABCD) = \sin \tfrac{1}{2} a \cdot \sin \tfrac{1}{2} c \div \sin \tfrac{1}{2} b \sin \tfrac{1}{2} d.$$

For if the perpendiculars from P on AB, BC, &c., be a, β, &c., and the staudtians of the triangles PAB, PBC, PCD, PDA being denoted by n_a, n_b, n_c, n_d, we have evidently

$$(P - ABCD) = \frac{n_a . n_c}{n_b . n_d} = \sin a \sin \alpha . \sin c \sin \gamma \div \sin b . \sin \beta . \sin d . \sin \delta$$

$$= \sin \tfrac{1}{2} a \sin \tfrac{1}{2} c \div \sin \tfrac{1}{2} b . \sin \tfrac{1}{2} d \text{ (equation (343)).} \quad (402)$$

3. If A, B, C, D be four points on a small circle X, the spherical triangle whose summits are the points of intersection of the arcs AB, CD; BC, DA, and CA, DB, is such that each side is the harmonic polar of the opposite vertex. This is called the harmonic triangle of the four points.

4. The harmonic polars of any point on the radical circle of two small circles with respect to these circles intersect on the radical circle.

5. If X, Y are two small circles, Z a great circle perpendicular to the great circle passing through the spherical centres of X, Y, the harmonic polars of any point of Z intersect on a great circle.

6. If a spherical quadrilateral be inscribed in a small circle (X), and at its angular points arcs of great circles be drawn touching X, their twelve points of intersection lie four by four on the sides of the harmonic triangle.

7. PASCAL'S THEOREM.—If a spherical hexagon be inscribed in a circle, the opposite sides intersect in pairs on a great circle.

8. A, B, C; A', B', C', are two triads of points on two great circles; prove that the intersections of the three pairs of arcs AB', $A'B$; BC', $B'C$; CA', $C'A$ lie on a great circle.

9. SALMON'S THEOREM.—Given any two points A and B and their harmonic polars, with respect to a small circle X, whose spherical centre is O. Let fall a perpendicular AP from A on the polar of B, and a perpendicular BQ from B on the polar of A; then, if A', B' be the harmonic conjugates of A, B, with respect to X, prove that

$$\cos OA : \cos OB :: \sin OA' \sin AP : \sin OB' \sin BQ.$$

10. BRIANCHON'S THEOREM.—If a spherical hexagon be described about a small circle X, the three arcs joining the opposite angular points are concurrent.

SECTION IV.—MUTUAL POWER OF TWO CIRCLES.

104. *If a, β, γ be the arcs connecting any point P to the vertices A, B, C of a trirectangular triangle, then $\cos^2 a + \cos^2 \beta + \cos^2 \gamma = 1$.*

DEM.—Since the triangles PAB, PAC are quadrantal, we have, from (equation (154),

$$\cos \beta = \sin a \cos BAP, \quad \cos \gamma = \sin a \cos CAP.$$

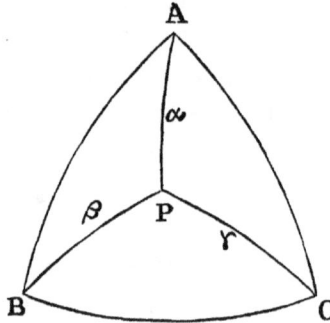

Fig. 43.

Hence

$$\cos^2 \beta + \cos^2 \gamma = \sin^2 a;$$

$$\therefore \quad \cos^2 a + \cos^2 \beta + \cos^2 \gamma = 1. \tag{403}$$

Cor.—If x, y, z be the perpendiculars from P on the sides of the triangle ABC; x, y, z are respectively equal $\cos a$, $\cos \beta$, $\cos \gamma$.

105. *If a, β, γ, a', β', γ be the angular distances (fig. 44) of two points P, P' from the vertices of the trirectangular triangle ABC; then $\cos PP' = \cos a \cos a' + \cos \beta \cos \beta' + \cos \gamma \cos \gamma'$,*

$\cos PP' = \cos a \cos a' + \sin a \sin a' \cos PAP' = \cos a \cos a'$

$\quad + \sin a \sin a' (\cos PAC \cos P'AC + \sin PAC \sin P'AC)$

$\quad = \cos a \cos a' + \cos \beta \cos \beta' + \cos \gamma \cos \gamma'$ (equation (154)).

$$\tag{404}$$

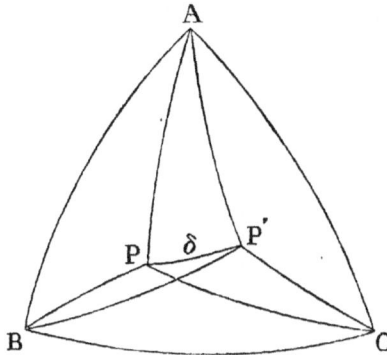

Fig. 44.

DEF. XXXVI.—*The product of the cosines of the spherical radii of two circles, subtracted from the cosine of the arc joining their spherical centres, gives a remainder, which is called the mutual power of the two circles.**

If the circles be denoted by letters with suffixes, we shall denote their mutual power by the suffixes. Thus the mutual power of the circles s_a, s_β shall be denoted by $a\beta$.

106. Frobenius's Theorem.—If $s_1, s_2, s_3, s_4, s_5; s_1', s_2', s_3', s_4', s_5'$ be any two systems of five circles on the sphere; then

$$\begin{vmatrix} 11', & 12', & 13', & 14', & 15' \\ 21', & 22', & 23', & 24', & 25' \\ 31', & 32', & 33', & 34', & 35' \\ 41', & 42', & 43', & 44', & 45' \\ 51', & 52', & 53', & 54', & 55' \end{vmatrix} = 0. \quad (405)$$

DEM.—Let x_1, y_1, z_1, &c., denote the normal co-ordinates of the centres of the circles, with respect to a fixed trirectangular

---

* The introduction of this term into Geometry is due to DARBOUX, "Annales de l'École Normale Supérieure," vol. I., 1872.

triangle, and r_1, r_2, &c., their spherical radii ; then, multiplying the determinants

$$\begin{vmatrix} 0, & x_1, & y_1, & z_1, & \cos r_1 \\ 0, & x_2, & y_2, & z_2, & \cos r_2 \\ 0, & x_3, & y_3, & z_3, & \cos r_3 \\ 0, & x_4, & y_4, & z_4, & \cos r_4 \\ 0, & x_5, & y_5, & z_5, & \cos r_5 \end{vmatrix} \begin{vmatrix} 0, & x_1', & y_1', & z_1', & -\cos r_1' \\ 0, & x_2', & y_2', & z_2', & -\cos r_2' \\ 0, & x_3', & y_3'. & z_3', & -\cos r_3' \\ 0, & x_4', & y_4', & z_4', & -\cos r_4' \\ 0, & x_5', & y_5', & z_5', & -\cos r_5' \end{vmatrix} ,$$

the proposition is evident.[*]

107. If the angle of intersection of the circles s_a, s_β be denoted by $\overline{a\beta}$, it follows at once, from equation (13), that the mutual power $(a\beta)$ is equal to $\sin r_a \cdot \sin r_\beta \cos \overline{a\beta}$. By this substitution, equation (405) is transformed into

$$\begin{vmatrix} \cos \overline{11'}, & \cos \overline{12'}, & \cos \overline{13'}, & \cos \overline{14'}, & \cos \overline{15'} \\ \cos \overline{21'}, & \cos \overline{22'}, & \cos \overline{23'}, & \cos \overline{24'}, & \cos \overline{25'} \\ \cos \overline{31'}, & \cos \overline{32'}, & \cos \overline{33'}, & \cos \overline{34'}, & \cos \overline{35'} \\ \cos \overline{41'}, & \cos \overline{42'}, & \cos \overline{43'}, & \cos \overline{44'}, & \cos \overline{45'} \\ \cos \overline{51'}, & \cos \overline{52'}, & \cos \overline{53'}, & \cos \overline{54'}, & \cos \overline{55'} \end{vmatrix} = 0. \quad (406)$$

108. If the second system of circles coincide with the first we have, for any system of five circles on the sphere,

$$\begin{vmatrix} 1, & \cos \overline{12}, & \cos \overline{13}, & \cos \overline{14}, & \cos \overline{15} \\ \cos \overline{21}, & 1, & \cos \overline{23}, & \cos \overline{24}, & \cos \overline{25} \\ \cos \overline{31}, & \cos \overline{32}, & 1, & \cos \overline{34}, & \cos \overline{35} \\ \cos \overline{41}, & \cos \overline{42}, & \cos \overline{43}, & 1, & \cos \overline{45} \\ \cos \overline{51}, & \cos \overline{52}, & \cos \overline{53}, & \cos \overline{54}, & 1 \end{vmatrix} = 0. \quad (407)$$

[*] This theorem is the fundamental one in a Memoir by HERR G. FRO-BENIUS, ''Anwendungen der Determinantentheorie auf die Geometrie des Maasses.'' CRELLE's *Journal*, Band 79, pp. 185–245, for the year 1875. It is also given in the *Philosophical Transactions*, vol. 177, part 2, for the year 1886, in a Memoir by R. LACHLIN, B.A., ''On Systems of Circles and Spheres.''

Cor. 1.—The condition that four circles should be cut orthogonally by a fifth is

$$\begin{vmatrix} 1, & \cos \overline{12}, & \cos \overline{13}, & \cos \overline{14} \\ \cos \overline{21}, & 1, & \cos \overline{23}, & \cos \overline{24} \\ \cos \overline{31}, & \cos \overline{32}, & 1, & \cos \overline{34} \\ \cos \overline{41}, & \cos \overline{42}, & \cos \overline{43}, & 1 \end{vmatrix} = 0. \quad (408)$$

For in this case $\cos \overline{15}$, $\cos \overline{25}$, &c., vanish.

Cor. 2.—The condition that four circles should be tangential to a fifth is

$$\begin{vmatrix} 0, & \sin^2\tfrac{1}{2} \overline{12}, & \sin^2\tfrac{1}{2} \overline{13}, & \sin^2\tfrac{1}{2} \overline{14} \\ \sin^2\tfrac{1}{2} \overline{21}, & 0, & \sin^2\tfrac{1}{2} \overline{23}, & \sin^2\tfrac{1}{2} \overline{24} \\ \sin^2\tfrac{1}{2} 31, & \sin^2\tfrac{1}{2} \overline{32}, & 0, & \sin^2\tfrac{1}{2} \overline{33} \\ \sin^2\tfrac{1}{2} \overline{41}, & \sin^2\tfrac{1}{2} \overline{42}, & \sin^2\tfrac{1}{2} \overline{43}, & 0 \end{vmatrix} = 0. \quad (409)$$

For if the circle s_5 touch each of the circles s_1, s_2, s_3, s_4, $\cos \overline{15}$, $\cos \overline{25}$, &c., become each equal to unity, and subtracting each of the four first columns from the last, we get the result just written.

109. If t_{12} be the arc of a great circle which is the common tangent of two small circles whose spherical radii are r_1, r_2, and angle of intersection ϕ_{12}, then it may be proved by equation (13) that the mutual power of the two circles is equal to $\sin r_1 \sin r_2 - 2 \cos r_1 \cos r_2 \sin^2\tfrac{1}{2}t_{12}$; and equating with the value $\sin r_1 \sin r_2 \cos \phi_{12}$ of § 107, we get

$$\sin^2\tfrac{1}{2} \phi_{12} = \sin^2\tfrac{1}{2} t_{12} \cdot \cot r_1 \cot r_2. \quad (410)$$

Hence in the determinant (409) the sines of half the angles of intersection of the circles s_1, s_2, s_3, s_4 may be replaced by the sines of half their common tangents, and denoting for shortness by $\overline{12}$ the sine of half the common tangent of the circles s_1, s_2, the condition is

$$\begin{vmatrix} 0, & \overline{12}^2, & \overline{13}^2, & \overline{14}^2 \\ \overline{21}^2, & 0, & \overline{23}^2, & \overline{24}^2 \\ \overline{31}^2, & \overline{32}^2, & 0, & \overline{34}^2 \\ \overline{41}^2, & \overline{42}^2, & \overline{43}^2, & 0 \end{vmatrix} = 0, \quad (411)$$

which, expanded, is equal to the product of the four factors

$$\overline{12}.\overline{34} \pm \overline{23}.\overline{14} \pm \overline{31}.\overline{24} = 0. \quad (412)$$

This theorem was first published in the *Proceedings of the Royal Irish Academy*, in a Paper by the author " On the Equations of Circles," in the year 1866.

110. If s_1, s_2, s_3, s_4 be a system of four great circles, and s_1', s_2', s_3', s_4' four other circles (great or small), then

$$\begin{vmatrix} 11', & 12', & 13', & 14' \\ 21', & 22', & 23', & 24' \\ 31', & 32', & 33', & 34' \\ 41', & 42', & 43', & 44' \end{vmatrix} = 0. \quad (413)$$

This is proved like Frobenius's theorem by multiplying the two determinants $(x_1, y_2, z_3, \cos r_4)$, $(x_1', y_2', z'_3, \cos r_4')$, the first of which vanishes ; since r_1, r_2, r_3, r_4, being the spherical radii of great circles, are each equal to a quadrant, and their cosines vanish.

111. If the second system in § 110 be great circles, and coincide with the first, we get, since the mutual power of two great circles is equal to the cosine of the arc joining their poles, a relation identical with equation (408) for the six arcs joining four points on a sphere $\overline{12}$, &c., denoting in this case the arcs joining the points 1, 2, &c. If the three first points be the vertices A, B, C of a spherical triangle, and the fourth any arbitrary point D, whose distances from A, B, C are denoted by α, β, γ, respectively, we get—

$$\begin{vmatrix} 1, & \cos c, & \cos b, & \cos \alpha \\ \cos c, & 1, & \cos a, & \cos \beta \\ \cos b, & \cos a, & 1, & \cos \gamma \\ \cos \alpha, & \cos \beta, & \cos \gamma, & 1 \end{vmatrix} = 0. \qquad (414)$$

112. If the first three circles of the second system in § 110 coincide with the first three circles of the first system, and the poles of these circles be the angular points of a spherical triangle ABC. Also, if s_4' be a great circle distinct from s_4, and the distances of the poles of these circles from the points A, B, C be α, β, γ; α', β', γ', respectively, and δ the arc joining their poles, we get—

$$\begin{vmatrix} 1, & \cos c, & \cos b, & \cos \alpha \\ \cos c, & 1, & \cos a, & \cos \beta \\ \cos b, & \cos a, & 1, & \cos \gamma \\ \cos \alpha', & \cos \beta', & \cos \gamma', & \cos \delta \end{vmatrix} = 0. \qquad (415)$$

EXERCISES.—XXX.

1. The incircles of a triangle and its colunar triangles have a fourth common tangential circle.—(HART.)

For if a, b, c be the sides of the original triangle, the direct common tangents of the incircles of the colunar triangles are $(b + c)$, $(c + a)$, $(a + b)$, respectively; and the transverse common tangents of the incircle of the original triangle and incircles of colunar triangles are $(b-c)$, $(c-a)$, $(a-b)$. Hence (see Art. 108) we have

$$\overline{23} = \sin\tfrac{1}{2}(b+c), \quad \overline{31} = \sin\tfrac{1}{2}(c+a), \quad \overline{12} = \sin\tfrac{1}{2}(a+b),$$

$$\overline{14} = \sin\tfrac{1}{2}(b+c), \quad \overline{24} = \sin\tfrac{1}{2}(c-a), \quad \overline{34} = \sin\tfrac{1}{2}(a+b).$$

Hence the condition (412) is fulfilled.

2. Prove that the mutual power of two circles is equal to the mutual power of two circles inverse to them.

3. If p, q, r be the normal co-ordinates of a point on the sphere, with respect to the sides of a spherical triangle ABC; prove that they are connected by the relation

$$
\begin{vmatrix}
-1, & \cos C, & \cos B, & p \\
\cos C, & -1, & \cos A, & q \\
\cos B, & \cos A, & -1, & r \\
p, & q, & r, & -1
\end{vmatrix} = 0 \qquad (416)
$$

In the equation (414), Art. 110, let the triangle ABC be replaced by its supplemental triangle, while the point D retains its position.

4. If a, b, c be the mutual distances of the spherical centres of three small circles whose radii are r_1, r_2, r_3; prove that if r be the radius of a circle cutting them orthogonally,

$$
4n^2 \sec^2 r =
\begin{vmatrix}
1, & \cos c, & \cos b, & \cos r_1 \\
\cos c, & 1, & \cos a, & \cos r_2 \\
\cos b, & \cos a, & 1, & \cos r_3 \\
\cos r_1, & \cos r_2, & \cos r_3, & 0
\end{vmatrix}.
$$

CHAPTER VII.

INVERSIONS.

SECTION I.—INVERSION IN SPACE.

113. DEF. XXXVII.—*Being given a fixed point S and a system of points A, B, C . . . . . if upon the right lines SA, SB, SC a system of points A', B', C' . . . . be determined by the relation SA . SA' = SB . SB' = SC . SC', &c. = constant, say k², the two systems A, B, C . . . A'B'C' . . . . are said to be inverse of each other. The point S is called the centre of inversion, and the sphere whose centre is S and radius k, the sphere of inversion.*

114. *The figure inverse to a plane is a sphere passing through the centre of inversion.*

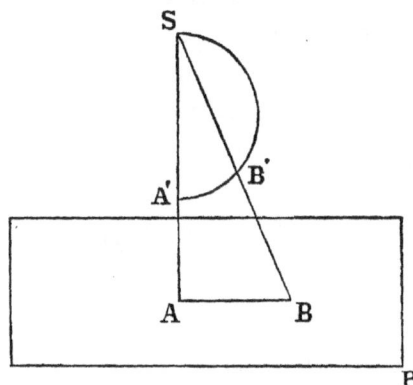

Fig. 45.

DEM.—From *S* draw the right line *SA* perpendicular to the plane *P*, and in *P* draw any line *AB* through *A*; then (*Sequel*, Prop. xx., p. 41), the inverse of the line *AB* is a circle *SA'B'* passing through *S*. Now, if the whole figure, consisting of the line AB and the circle $SA'B'$, turn round the line SA, the line AB will describe the plane P, and the circle $SA'B'$ will describe a sphere, which is the inverse of the plane.

Cor. 1.—The inverse of a sphere passing through the centre of inversion is a plane.

Cor. 2.—If m', n' be the inverses of the points m, n, then

$$mn = \frac{k^2 m'n'}{Sm' \cdot Sn'}. \tag{417}$$

This follows from the triangles Smn, $Sm'n'$, which are evidently similar.

115. *The inverse of a sphere which does not pass through the centre of inversion is a sphere.*

Dem.—If X be any circle coplanar with S, its inverse will be another circle X', coplanar with S and X, and S will be the centre of similitude of the two circles (*Sequel*, Prop. I., p. 95); then, if the figure consisting of the two circles be turned round the line through the centres of both circles, the spheres described will be inverse to each other with respect to the point S.

Cor. 1—The inverse of a circle with respect to any point in space is another circle.

For the first circle may be regarded as the curve of intersection of two spheres; its inverse will be the curve of intersection of the inverse spheres.

Cor. 2.—The cone which has for base a small circle of the sphere, and vertex any point, cuts the sphere again in another circle.

EXERCISES.—XXXI.

1. The locus of a point, from which two unequal spheres can be inverted into two equal spheres, is a sphere.

2. The locus of a point, from which three unequal spheres can be inverted into three equal spheres, is a circle.

3. Dupuis' Theorem.—If a variable sphere touch three fixed spheres, the locus of its point of contact with each fixed sphere is a circle.

For if a variable sphere be inscribed in a trihedral angle, the locus of its point of contact on each face of the trihedral is a right line; and when we invert, the planes become spheres, and the right lines circles.

4. Prove that four unequal spheres can be inverted into four equal spheres.

Section II.—Stereographic Projection.

116. Def. XXXVIII.—*Stereographic projection is the drawing of the circles of the sphere upon the plane of one of its great circles (called the plane of the primitive) by lines drawn from the pole of that great circle to all the points of the circles to be projected.*

It is evident that the plane is the projection in space of the sphere, the value of the constant k^2 being $2R^2$.

117. *The stereographic projection of any circle is a circle.*

This follows at once from § 115, *Cor.* 1, but we here give an independent proof.

Dem.—1°. *In the case of a small circle.*

Let *Pm* be a generator of the cone, touching the sphere along

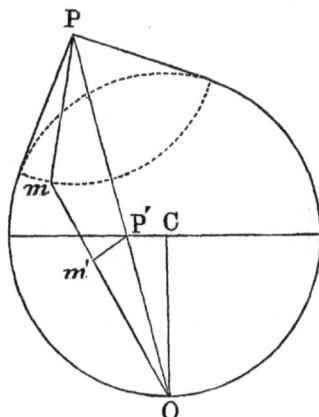

Fig. 46.

the given circle, and let *P'*, *m'* be the projections of the summit of the cone, and of the point m of the circle, on the plane of the primitive. Let O be the pole of the primitive, and let Pm produced meet a tangent plane to the sphere at O in T; then, since the plane of the primitive and the tangent plane at O are parallel, the plane OmP cuts them in parallel lines (Euc. XI., XVI.). Hence the angle $P'm'O$ is equal to $m'OT$; but the angle $m'OT$ is equal to OmT, since the tangents mT, OT are equal. Hence the angle $P'm'O$ is equal to the supplement of PmO, and the angle O is common to the two triangles PmO, $P'm'O$; therefore $OP : Pm :: OP' : P'm'$; and since the three first terms of this proportion are given, the fourth, $P'm'$, is given. *Hence the locus of m' is a circle whose centre is collinear with the pole of the primitive and the vertex of the cone.*—(CHASLES.)

Cor.—If the cones circumscribed to a sphere along a system of circles have their vertices in a line passing through the pole of the primitive, their stereographic projections is a system of concentric circles.

2°. *In the case of a great circle.*

Let A be the pole of the primitive, and let the plane of the circle AIB be perpendicular to the line of intersection of the

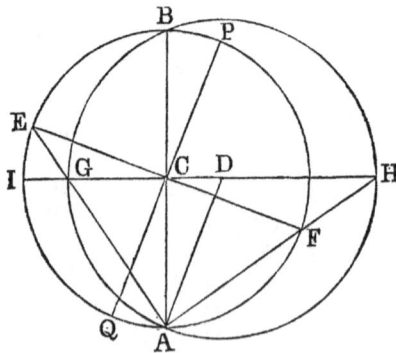

Fig. 47.

plane of the primitive with the plane of the circle to be projected, and let it intersect the plane of that circle in the line

EF. Let *IC* be perpendicular to *AB.* Join *AE, AF,* intersecting *IC* in the points *G, H.* Now, since *GAH* is a right
angle, and *AC* is perpendicular to *GH,* the angle *AHG = GAC*
(Euc. VI. viii.) = *AEC* (Euc. I. v.). Hence the triangles *EAF*
and *HAG* are inversely similar, and therefore the section made
by the plane of the primitive with the cone, whose vertex is *A,*
and which stands on the great circle *EF,* is an antiparallel
section. Hence it is a circle.

Cor. 1.—The projections of the poles *P, Q* of the great circle
EF will be inverse points with respect to its projection.

Cor. 2.—If the plane of a small circle be parallel to the
plane of *EF,* the projection of *P* and *Q* will be inverse points
with respect to its projection.

Cor. 3.—A system of small circles, whose planes are parallel,
will project into a system of coaxal circles.

Cor. 4.—Every circle whose plane passes through the pole
of the primitive is projected into a right line.

118. *The angle made by any two circles on the sphere is equal
to the angle made by their projections on the plane of the primitive.*

Dem.—Let *O* be the pole of the primitive, *M* the point in
which the circles intersect; and let *MT, MV* the tangents to
the circles at *M,* meet the tangent plane to the sphere at *O* in
the points *T, V.* Join *OT, OV, TV;* then evidently the angle
TMV = TOV; but since the tangent plane at *O* is parallel to
the plane of the primitive, the lines *OT, OV* are parallel to
the projections of the lines *MT, MV.* Hence the angle *TOV* is
equal to the angle between the projections.

Cor. 1.—Any circle whose plane is perpendicular to the
plane of the primitive is projected into a circle orthogonal to
the primitive.

Cor. 2.—A system of coaxal circles on the sphere is projected
into a system of coaxal circles on the plane of the primitive.

For a system of coaxal circles on the sphere is intersected orthogonally by a system of circles passing through the two limiting points (§ 94). Hence the projections are intersected orthogonally by a system of circles passing through two points.

119. *Applications to Spherical Trigonometry.*

Let ABC be a spherical triangle, $AB'C$ a colunar triangle; then, if O, the pole of the primitive, be the antipodes of A, the sides AB, AC, AB' will project into right lines ab, ac, ab', and the circle BCB' into the circle bcb'. Join bc, cb'; then the angles of the figure formed by the lines ab, ac, and the arc bc, are respectively equal to the angles of the spherical triangle ABC (Art. 117); but the sum of the angles of the rectilineal triangle abc is two right angles, hence the sum of the angles formed by the arc bc with its chord is equal to the spherical excess $2E$; therefore one of them is equal to E.

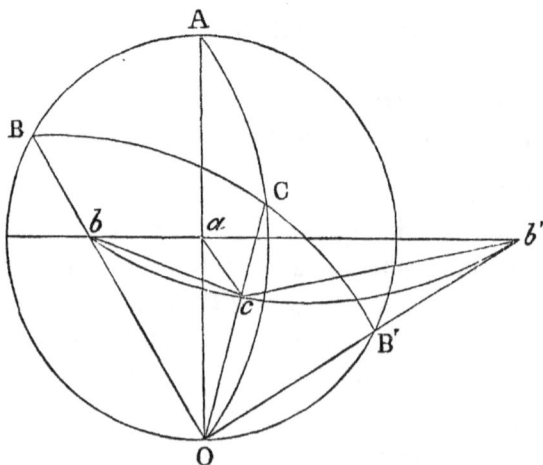

Fig. 48.

Or thus:—If a circle be described about the triangle abc, the angle made by this circle with the arc bc is (§ 118) equal to the angle made by the circumcircle of the triangle ABC with the side BC, and this is equal to $(A - E)$ (Exercises XXIII. 15);

and the angle made by the circumcircle of abc with the chord bc is equal to A (Euc. III. xxxii.). Hence the angle between the arc bc and its chord is equal to E.

Cor. 1.—If the radius of the sphere be unity, the sides of the rectilineal triangle can be expressed in terms of the spherical triangle. Thus, evidently,

$$ab = \tan AOB = \tan \tfrac{1}{2} AB = \tan \tfrac{1}{2} c. \qquad (418)$$

$$ac = \tan AOC = \tan \tfrac{1}{2} AC = \tan \tfrac{1}{2} b. \qquad (419)$$

Again, $\quad bc : ab : : \sin bac : \sin acb : : \sin A : \sin (C - E)$;

$$\therefore \quad bc : \tan \tfrac{1}{2} c : : \frac{n}{\sin b \,.\, \sin c} : \frac{n}{2 \sin \tfrac{1}{2} a \,.\, \sin \tfrac{1}{2} b \,.\, \cos \tfrac{1}{2} c} ;$$

$$\therefore \quad bc = \frac{\sin \tfrac{1}{2} a}{\cos \tfrac{1}{2} b \, \cos \tfrac{1}{2} c}. \qquad (420)$$

Similarly,
$$b'c = \frac{\cos \tfrac{1}{2} a}{\cos \tfrac{1}{2} b \, \sin \tfrac{1}{2} c}. \qquad (421)$$

The equation (420) may be got from equation (417) by putting $k^2 = 2$. Thus—

$$bc = \frac{2 \text{ chord } BC}{OB \,.\, OC} = \frac{\sin \tfrac{1}{2} a}{\cos \tfrac{1}{2} b \, \cos \tfrac{1}{2} c} ;$$

and (421) from (420), by the substitution of § 78.

Cor. 2.—If the angles of the rectilineal triangle abc be denoted by α, β, γ, we have

$$\alpha = A, \quad \beta = B - E, \quad \gamma = C - E. \qquad (422)$$

Exercises.—XXXII.

1. Prove the fundamental formula (13) by stereographic projection. From the triangle abc we have

$$(bc)^2 = (ca)^2 + (ab)^2 - 2 (ca) (ab) \cos A,$$

and substitute from equations (418)–(420).

2. Prove Napier's Analogies.

We have
$$\frac{\tan \tfrac{1}{2} (\beta - \gamma)}{\tan \tfrac{1}{2} a} = \frac{ac - ab}{ac + ab}.$$

And substitute from equations (418)–(422).

3. Prove Delambre's Analogies.

From abc we have $bc = \dfrac{(ac + ab)\,\sin\frac{1}{2}\,\alpha}{\cos\frac{1}{2}\,(\beta - \gamma)} = \dfrac{(ac - ab)\,\cos\frac{1}{2}\,\alpha}{\sin\frac{1}{2}\,(\beta - \gamma)},$

and substitute as before.

4. Being given three circles on the sphere; there are eight points on the sphere, any one of which, if taken as the pole of the primitive, the three circles will be projected into three equal circles.—(STEINER.)

5. $\tan\frac{1}{2}\,E = \sqrt{\tan\frac{1}{2}\,s\,.\,\tan\frac{1}{2}\,(s - a)\,\tan\frac{1}{2}\,(s - b)\,\tan\frac{1}{2}\,(s - c)}.$

Express $\tan\frac{1}{2}$ the angle $ab'c$, in terms of the sides of $ab'c$.

6. Prove that $\cos E = \dfrac{1 + \cos a + \cos b + \cos c}{4\,\cos\frac{1}{2}\,a\,\cos\frac{1}{2}\,b\,\cos\frac{1}{2}\,c}.$

In the triangle $ab'c$, we have $ab' = ac\,\cos cab + cb'\,\cos cb'a$.

7. To express the spherical excess of a spherical quadrilateral in terms of its sides and diagonals.

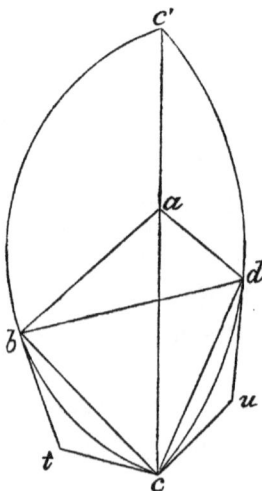

Fig. 49.

Let $ABCD$ be the quadrilateral; and denoting the sides and diagonals AB, BC, CD, DA, AC, BD by a, b, c, d, e, f, respectively; then taking the antipodes of A for the pole of the primitive, the arcs AB, AD, AC will project into right lines ab, ac, ad; and the arcs BC, CD into arcs of circles

bc, cd; then drawing bt, ct tangents to bc, and cu du, tangents to cd, we have in the plane hexagon $abtcud$ the sum of the angles

$$A + B + C + D + t + u = 4\pi; \quad \text{but } A + B + C + D = 2\pi + 2E;$$

$$\therefore \quad 2E = 2\pi - t - u.$$

Again, if the circles bc, cd intersect again in c', c' is the stereographic projection of the antipodes. Hence the right line ca produced will pass through c. Join bc, bc'; dc, dc', then the angle $tbc = bc'c$. Hence $bc'c$ is half the supplement of t, and $cc'd$ half the supplement of u;

$$\therefore \quad 2bc'd + t + u = 2\pi; \quad \therefore \quad bc'd = E.$$

Now from the plane triangle $bc'd$, we have

$$\sin^2 \tfrac{1}{2} bc'd = \sin^2 \tfrac{1}{2} E = \frac{(bc' + bd - c'd)(bd + c'd - bc')}{4bc' . dc'};$$

but

$$bd = \frac{\sin \tfrac{1}{2} f}{\cos \tfrac{1}{2} a \cos \tfrac{1}{2} d}, \quad bc' = \frac{\cos \tfrac{1}{2} b}{\cos \tfrac{1}{2} a \sin \tfrac{1}{2} c}, \quad dc' = \frac{\cos \tfrac{1}{2} c}{\cos \tfrac{1}{2} d \sin \tfrac{1}{2} e};$$

$$\therefore \quad \sin^2 \tfrac{1}{2} E =$$

$$\frac{(\sin \tfrac{1}{2} e . \sin \tfrac{1}{2} f + \cos \tfrac{1}{2} a \cos \tfrac{1}{2} c - \cos \tfrac{1}{2} b \cos \tfrac{1}{2} d)(\sin \tfrac{1}{2} e \sin \tfrac{1}{2} f - \cos \tfrac{1}{2} a \cos \tfrac{1}{2} c + \cos \tfrac{1}{2} b \cos \tfrac{1}{2} d)}{4 \cos \tfrac{1}{2} a \cos \tfrac{1}{2} b \cos \tfrac{1}{2} c \cos \tfrac{1}{2} d}$$

8. If a spherical quadrilateral be cyclic, prove that

$$\sin^2 \tfrac{1}{2} E = \frac{\sin \tfrac{1}{2}(s - a) \sin \tfrac{1}{2}(s - b) \sin \tfrac{1}{2}(s - c) \sin \tfrac{1}{2}(s - d)}{\cos \dfrac{a}{2} \cos \dfrac{b}{2} \cos \dfrac{c}{2} \cos \dfrac{d}{2}}. \quad (424)$$

9. If the cyclic quadrilateral be circumscribed to another circle, prove

$$\sin^2 \tfrac{1}{2} E = \tan \tfrac{1}{2} a . \tan \tfrac{1}{2} b \tan \tfrac{1}{2} c . \tan \tfrac{1}{2} d. \quad (425)$$

10. Being given four circles in a plane, prove that the plane can be inverted into a sphere, so that the four circles on the plane will be the stereographic projections of four equal circles on the sphere.—(STEINER.)

11. If A', B', C' be the stereographic projections of the angular points of the spherical triangle ABC; and if the angles of the plane triangle $A'B'C'$ be respectively equal to those of the spherical triangle, each diminished by one-third of the spherical excess, prove that the arcs drawn from A, B, C to one of the poles of the primitive divides the area of ABC into three equal parts.

Observation.—The applications of Stereographic Projection to Spherical Trigonometry, contained in § 119 and in Exercises XXXII., are taken from M. PAUL SERRET, *Méthodes des Géométrie*, pp. 30–44.

CHAPTER VIII.

POLYHEDRA.

SECTION I.—REGULAR POLYHEDRA.

120. *If S be the number of solid angles, F the number of faces, E the number of edges of any polyhedron* (see Appendix to EUCLID, 6th edition, p. 283), $S + F = E + 2.$—(EULER.)

DEM.—With any point in the interior of the polyhedron as centre, describe a sphere of radius r, and draw lines from the centre to each solid angle; let the points in which these meet the surface of the sphere be joined by arcs of great circles. These arcs will divide the surface into F spherical polygons. Now if s denote the sum of the angles of any of these polygons, and m the number of its sides, its area is $r^2(s - (m-2)\pi)$, but the sum of the areas of all the polygons is equal to the surface of the sphere or $4\pi r^2$. Hence, since there are F polygons, we have $4\pi = \Sigma s - \pi\Sigma m + 2F\pi$; but Σs is evidently equal to $2\pi S$, and Σm is the number of the sides of all the polygons, and therefore equal to $2E$. Hence $4\pi = 2\pi S - 2E\pi + 2F\pi$;

$$\therefore \quad S + F = E + 2. \tag{426}$$

121. *There can be only five regular polyhedra.*

DEM.—Let m be the number of sides in each face, and n the number of plane angles in each solid angle, then the entire number of plane angles is equal to mF or nS or $2E$. Hence we have the equations

$$mF = nS = 2E \quad \text{and} \quad S + F = E + 2.$$

Therefore solving for S, F, and E, we get

$$S = \frac{4m}{2(m+n)-mn}, \quad F = \frac{4n}{2(m+n)-mn}, \quad E = \frac{2mn}{2(m+n)-mn}. \tag{427}$$

Since the denominator in these expressions must be positive, $\dfrac{1}{m} + \dfrac{1}{n}$ must be greater than $\dfrac{1}{2}$; but n cannot be less than 3, since a solid angle cannot be formed by less than three plane angles. Hence m cannot be greater than 5. The following will be found to be the only admissible system of values for m and n, viz.,

$$3, 3; \quad 4, 3; \quad 3, 4; \quad 5, 3; \quad 3, 5;$$

and the corresponding polyhedra are the *Tetrahedron, Cube, Octahedron, Dodecahedron,* and *Icosahedron,* or solids of 4, 6, 8, 12, 20 faces, which have respectively 4, 8, 6, 20, 12 vertices.

122. *If I denote the inclination of two adjacent faces of a regular polyhedron,*

$$sin\ \tfrac{1}{2} I = cosec\ \frac{\pi}{m} \cdot cos\ \frac{\pi}{n}. \tag{428}$$

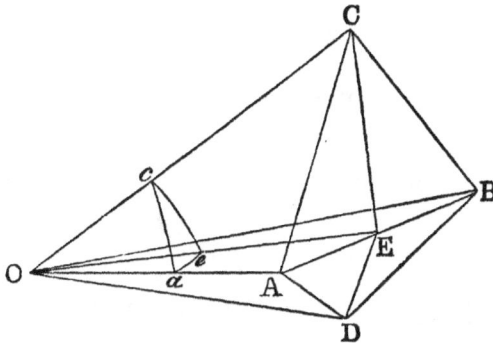

Fig. 50.

DEM.—Let AB be the side common to the two faces, C and D their centres, from which let CE, DE be drawn perpendicular to AB; then the angle between CE and ED will be equal to I. In the plane of the lines CE, DE, let CO and DO be drawn at right angles to them, and meeting in O. Join OA, OE, OB, and from O as centre suppose a sphere to be

K

described, cutting OA, OE, OC in the points a, e, c, respectively; then aec is a spherical triangle, having the angle e right; also $cae = \dfrac{\pi}{n}$ and $ace = \dfrac{\pi}{m}$, and by equation (111), $\sin ace = \cos cae \div \cos ce$; but $\cos ce = \cos coe = \sin \frac{1}{2} I$;

$$\therefore \quad \sin \tfrac{1}{2} I = \operatorname{cosec} \frac{\pi}{m} \cos \frac{\pi}{n}.$$

Cor. 1.—The following are the values of I for the five regular *polyhedra.* Thus, denoting them by P_4, P_6, P_8, P_{12}, P_{20} :—

In P_4, $\cos I = \dfrac{1}{3}$; in P_6, $I = \dfrac{\pi}{2}$; in P_8, $\cos I = -\dfrac{1}{3}$;

in P_{12}, $\cos I = \dfrac{-1}{\sqrt{5}}$; in P_{20}, $\cos I = -\dfrac{1}{3}\sqrt{5}$.

Cor. 2.—If r be the radius of the inscribed sphere, and a a side of one of the faces,

$$r = \frac{a}{2} \cot \frac{\pi}{m}.\tan\frac{I}{2}. \tag{429}$$

For $r = CE.\tan CEO = CE \tan \dfrac{I}{2} = \dfrac{a}{2}\cot\dfrac{\pi}{m}\tan\dfrac{I}{2}.$

Cor. 3.—If R be the circumradius of the polyhedron,

$$R = \frac{a}{2}\tan\frac{\pi}{n}\tan\frac{I}{2}. \tag{430}$$

Cor. 4.—The surface of a regular polyhedron

$$= \frac{ma^2 F}{4}\cot\frac{\pi}{m}. \tag{431}$$

Cor. 5.—The volume of a regular polyhedron

$$= \frac{ma^2 r F}{12}\cot\frac{\pi}{m}. \tag{432}$$

Cor. 6.—The octahedron is the reciprocal of the cube, and the icosahedron of the dodecahedron.

Exercises.—XXXIII.

1. Find the values of E, F, S for each of the regular polyhedra.

2. Prove that the centres of the faces of the polyhedra P_4, P_6, P_8, P_{12}, P_{20} are respectively the summits of polyhedra P_4, P_8, P_6, P_{20}, P_{12}.

3. Find the ratios between the volumes of a tetrahedron or cube and the volume of the solid, whose summits are the centres of its faces.

4. Prove that the inradius of $P_4 =$ three times its circumradius.

5. In the same case, the radius of the sphere touching its six edges is a mean proportional between the inradius and circumradius.

6. Prove that the ratio of the inradius to circumradius is the same in P_6 and P_8, and also in P_{12} and P_{20}.

7. In any convex polyhedron (regular or irregular), prove that the number of faces having an odd number of sides is even, and that the number of solid angles having an odd number of edges is uneven.

8. In every convex polyhedron, the number of triangular faces increased by the number of trihedral angles is equal to or greater than eight.

9. Every convex polyhedron must have either triangular, or quadrangular, or pentagonal faces, and trihedral, or tetrahedral, or pentrahedral angles.*

Section II.—Parallelopipeds and Tetrahedra.

123. *To find the volume of a parallelopiped in terms of three conterminous edges and their inclinations.*

Let OA, OB, OC be the three edges, and let their lengths be a, b, c, respectively; and let the angles BOC, COA, AOB be denoted by a, β, γ. Draw AD perpendicular to the plane BOC, and describe a sphere, with O as centre, meeting the lines OA, OB, OC, OD in the points a, b, c, d, respectively. The

---

* The most important recent works which treat of the polyhedra are Allman's "Greek Geometry from Thales to Euclid," and "Lectures on the Icosahedron," by Professor Klein, Göttingen. This is a very remarkable work, showing the great importance of the polyhedra in the higher departments of modern Analysis.

volume of the parallelopiped is equal to the product of the base and altitude $= bc \sin a \cdot AD$; but $AD = a \cdot \sin AOD = a \cdot \sin aod$;

$$\therefore \quad \text{vol.} = abc \sin a \cdot \sin aod = 2abc \cdot n, \qquad (433)$$

n being the first staudtian of the solid angle $O - ABC$

$$= abc \sqrt{1 - \cos^2 a - \cos^2 \beta - \cos^2 \gamma + 2 \cos a \cos \beta \cos \gamma}. \qquad (434)$$

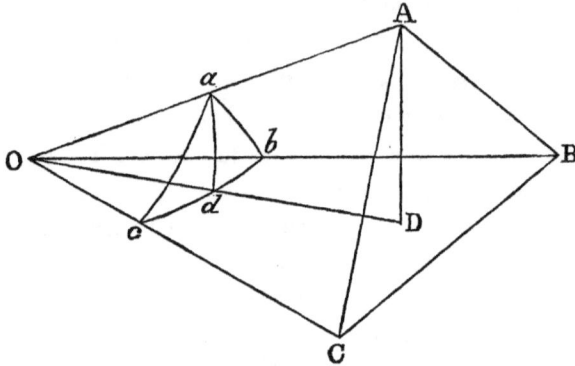

Fig. 51.

Cor. 1.—The volume of the tetrahedron $O - ABC = \frac{1}{3} abcn.$
$$(435)$$

Cor. 2.—To find the volume of a tetrahedron in terms of its six edges. Let $BC = a'$, $CA = b'$, $AB = c'$; then we have

$$\cos a = \frac{b^2 + c^2 - a'^2}{2bc}, \quad \cos \beta = \frac{c^2 + a^2 - b'^2}{2ca}, \quad \cos \gamma = \frac{a^2 + b^2 - c'^2}{2ab},$$

$$V = \frac{1}{6} abc \sqrt{1 - \cos^2 a - \cos^2 \beta - \cos^2 \gamma + 2 \cos a \cos \beta \cos \gamma}.$$

Hence $\quad 144 V^2 = \Sigma a^2 a'^2 (b^2 + b'^2 + c^2 + c'^2 - a^2 - a'^2)$

$$- a^2 b'^2 c'^2 - b^2 c'^2 a'^2 - c^2 a'^2 b'^2 - a^2 b^2 c^2. \qquad (436)$$

Cor. 3.—If T_a, T_b, T_c be the areas of the three faces OBC, OCA, OAB of the tetrahedron $O - ABC$, and N the second staudtian of the solid angle $O - ABC$,

$$V^2 = \frac{4}{9} T_a T_b T_c N. \qquad (437)$$

For $\quad T_a = \frac{1}{2} bc \sin a, \;\; T_b = \frac{1}{2} ca \sin \beta, \;\; T_c = \frac{1}{2} ab \sin \gamma,$

$$N = \frac{2n^2}{\sin a . \sin \beta . \sin \gamma}.$$

Hence $\quad\quad \frac{4}{9} T_a T_b T_c N = \frac{1}{9} a^2 b^2 c^2 n^2 = V^2.$

Cor. 4.—The second staudtians of the solid angles of a tetrahedron are proportional to the areas of the opposite faces.

This follows at once from *Cor.* 3.

Cor. 5.—The volume of a tetrahedron is equal to $\frac{2}{3}$ of the product of the areas of two faces by the sine of their dihedral angle divided by the length of their common edge.

For the vol. $= \frac{1}{3}$ of the triangle $OBC . AD$, and $AD = 2$ triangle ABC, multiplied by sine of the dihedral angle of the faces OBC, ABC divided by BC.

Cor. 6.—The products of opposite edges of a tetrahedron are proportional to the products of the sines of the corresponding dihedral angles.

124. If a, β, γ, δ denote the areas of the four faces of a tetrahedron,

$$a^2 = \beta^2 + \gamma^2 + \delta^2 + 2\beta\gamma \cos(\widehat{\beta\gamma}) + 2\gamma\delta \cos(\widehat{\gamma\delta}) + 2\delta\beta \cos(\widehat{\delta\beta}).$$

$$(438)$$

Dem.—We have

$$a = \beta \cos(a\beta) + \gamma \cos(a\gamma) + \delta \cos(a\delta). \quad (1)$$

$$\beta = a \cos(\beta a) + \gamma \cos(\beta\gamma) + \delta \cos(\beta\delta). \quad (2)$$

$$\gamma = a \cos(\gamma a) + \beta \cos(\gamma\beta) + \delta \cos(\gamma\delta). \quad (3)$$

$$\delta = a \cos(\delta a) + \beta \cos(\delta\beta) + \gamma \cos(\delta\gamma). \quad (4)$$

Multiplying by a, β, γ, δ, respectively, and subtracting the sum of the three last products from the first, we get the above result.

Cor.—If we eliminate a, β, γ, δ from the equations (1), (2), (3), (4), we get the following relation between the six dihedral angles of a tetrahedron :—

$$\begin{vmatrix} -1, & \cos a\beta, & \cos a\gamma, & \cos a\delta \\ \cos \beta a, & -1, & \cos \beta\gamma, & \cos \beta\delta \\ \cos \gamma a, & \cos \gamma\beta, & -1, & \cos \gamma\delta \\ \cos \delta a, & \cos \delta\beta, & \cos \delta\gamma, & -1 \end{vmatrix} = 0. \quad (439)$$

This relation may also be easily inferred from equation (414.)

125. *To find the diagonal of a parallelopiped in terms of three conterminous edges and their inclinations.*

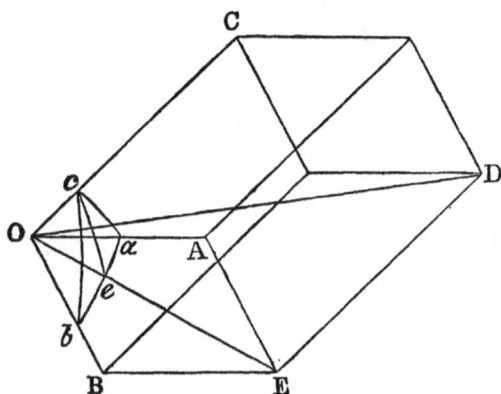

Fig. 52.

Let the edges OA, OB, OC be denoted by a, b, c; and the angles OBC, OCA, OAB by a, β, γ, respectively; let OD be the diagonal required; and OE the diagonal of the face OAB; then

$$OD^2 = OE^2 + ED^2 + 2OE \cdot ED \cdot \cos COE$$

$$= a^2 + b^2 + 2ab \cos \gamma + c^2 + 2c \cdot OE \cdot \cos COE.$$

Describe a sphere, with O as centre, cutting OA, OB, OC, OE

in the points a, b, c, e, respectively; then we have, by Stewart's theorem,

$$\cos COE = (\cos a . \sin aOe + \cos \beta \sin bOe) \div \sin \gamma;$$

$$\therefore \quad OD^2 = a^2 + b^2 + c^2 + 2ab \cos \gamma$$

$$+ \frac{2c . OE}{\sin \gamma} (\cos a . \sin aOe + \cos \beta \sin bOe);$$

but $\quad OE \sin aOe = b \sin \gamma$, and $OE . \sin bOe = a \sin \gamma$.

Hence

$$OD^2 = a^2 + b^2 + c^2 + 2bc \cos a + 2ca \cos \beta + 2ab \cos \gamma. \quad (440)$$

126. *To find the radius of a sphere circumscribed to a tetrahedron.*

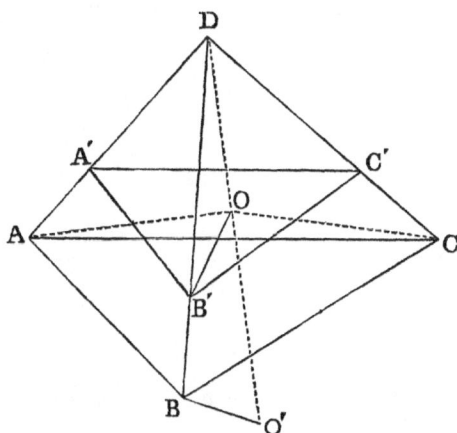

Fig. 53.

FIRST METHOD (STAUDT'S):—

Through the centre O of the circumsphere draw a plane perpendicular to the radius DO, cutting the sides of the trihedral angle D in the plane triangle $A'B'C'$. This plane being parallel to the tangent plane to the sphere at D,

$A'B'$ is antiparallel to AB in the angle ADB,

$B'C'$,, BC ,, BDC,

$C'A'$,, CA ,, CDA.

Let $DA = a$, $DB = b$, $DC = c$, $BC = a'$, $CA = b'$, $AB = c'$,

$$DA' = \alpha, \quad DB' = \beta, \quad DC' = \gamma, \quad B'C' = \alpha', \quad C'A' = \beta', \quad A'B' = \gamma',$$

we have

$$\frac{\alpha'}{a'} = \frac{\beta}{c} = \frac{\gamma}{b}; \quad \therefore \quad \alpha' = \frac{a'\beta}{c} = \frac{a'\beta b}{bc} = \frac{a' \cdot a\alpha}{bc} = a a' \cdot \frac{a\alpha}{abc};$$

$$\beta' = bb' \cdot \frac{b\beta}{abc}, \quad \gamma' = cc' \cdot \frac{c\gamma}{abc}.$$

If O' be the second extremity of the diameter DO' of the sphere, we have

$$a\alpha = b\beta = c\gamma = 2R^2 = DO \cdot DO'.$$

Hence $\quad \alpha' = \dfrac{aa' \cdot 2R^2}{abc}, \quad \beta' = \dfrac{bb' \cdot 2R^2}{abc}, \quad \gamma' = \dfrac{cc' \cdot 2R^2}{abc}.$

Putting $aa' = a_1$, $bb' = b_1$, $cc' = c_1$, and $a_1 + b_1 + c_1 = 2s_1$, we have triangle

$$A'B'C' = \frac{4R^4}{a^2 b^2 c^2} \sqrt{s_1 \cdot (s_1 - a_1)(s_1 - b_1)(s_1 - c_1)} = \frac{4R^4}{a^2 b^2 c^2} S_1.$$

We have also

$$\frac{DABC}{DA'B'C'} = \frac{abc}{\alpha\beta\gamma} = \frac{a^2 b^2 c^2}{a\alpha \cdot b\beta \cdot c\gamma} = \frac{a^2 b^2 c^2}{8R^6};$$

then

$$\frac{DABC}{\dfrac{4R^4}{a^2 b^2 c^2} S_1 \cdot \frac{1}{3} R} = \frac{a^2 b^2 c^2}{8R^6}; \quad \therefore \quad R = \frac{S_1}{6 \, DABC}. \qquad (441)$$

SECOND METHOD (DOSTOR, *Nouvelles Annales*, 1874, p. 523):—

Let $DABC$ be the tetrahedron, O its circumcentre; A', B', C' the middle points of the edges DA, DB, DC, which, as before, are denoted by a, b, c, respectively. Draw OM parallel to DA, meeting the face BDC in M, and MN parallel to BD. Now, since the projection of DO, and of DN, NM, MO on DA, DB, DC, DO are equal, we have

$$\tfrac{1}{2}\,a = DN \cos ac + NM \cos ab + MO.$$

$$\tfrac{1}{2}\,b = DN \cos bc + NM + MO \cos ab,$$

$$\tfrac{1}{2}\,c = DN + NM \cos bc + MO \cos ac,$$

$$R \;= DN \cos (c,\,R) + NM \cos (b,\,R) + MO \cos (a,\,R).$$

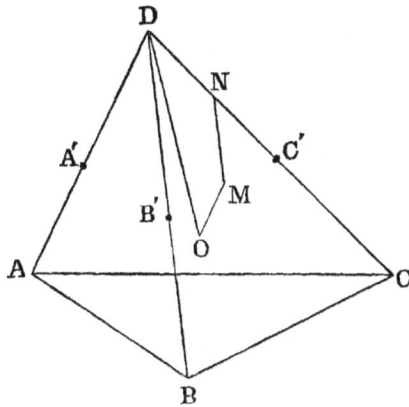

Fig. 54.

Hence $\quad 2R^2 = MO . a + MN . b + DN . c.$

$$a = 2MO + 2MN \cos ab + 2DN \cos ac.$$

$$b = 2MO \cos ab + 2MN + 2DN \cos bc.$$

$$c = 2MO \cos ca + 2MN \cos cb + 2DN.$$

Hence, eliminating MO, MN, DN, we get

$$\begin{vmatrix} 2R^2, & a, & b, & c \\ a, & 2, & 2\cos ab, & 2\cos ac \\ b, & 2\cos ba, & 2, & 2\cos bc \\ c, & 2\cos ca, & 2\cos cb, & 2 \end{vmatrix} = 0.$$

Now n being the first staudtian of the solid angle $D - ABC$, we have $4n^2 =$

$$\begin{vmatrix} 1, & \cos ab, & \cos ac \\ \cos ba, & 1, & \cos bc \\ \cos ca, & \cos bc, & 1 \end{vmatrix}.$$

Therefore $64R^2n^2 =$

$$\begin{vmatrix} 0, & a, & b, & c \\ a, & 2, & 2\cos ab, & 2\cos ac \\ b, & 2\cos ba, & 2, & 2\cos bc \\ c, & 2\cos ca, & 2\cos cb, & 2 \end{vmatrix} = \frac{1}{a^2b^2c^2} \begin{vmatrix} 0, & a^2, & b^2, & c^2 \\ a^2, & 2a^2, & 2ab\cos ab, & 2ac\cos ac \\ b^2, & 2ba\cos ba, & 2b^2, & 2bc\cos bc \\ c^2, & 2ca\cos ca, & 2cb\cos cb, & 2c^2 \end{vmatrix}$$

Hence $64a^2b^2c^2n^2R^2 =$

$$\begin{vmatrix} 0, & a^2, & b^2, & c^2 \\ a^2, & 0, & c'^2, & b'^2 \\ b^2, & c'^2, & 0, & a'^2 \\ c^2, & b'^2, & a'^2 & 0 \end{vmatrix} = -\Sigma(aa')^4 + 2\Sigma(aba'b')^2. \qquad (442)$$

Cor.—$24VR = \{2\Sigma(aba'b')^2 - \Sigma(aa')^4\}^{\frac{1}{2}}$.

127. The Isosceles Tetrahedron.—(Neuberg.)

Def. XXXIX.—*An isosceles tetrahedron is one whose opposite edges are equal.*

From the definition it follows at once (Euc. I., viii. xxxii.) that the four faces are equal, and that the sum of the plane angles forming each trihedral angle is equal to two **right** angles.

128. If we suppose $BC = AD = \alpha$, $AC = DB = \beta$, $AB = DC = \gamma$; then denoting by a, b, c the angles of the triangle ABC, they are also the face angles of the trihedral angle $D - ABC$; and representing by A, B, C the dihedrals DA, DB, DC, we have (§ 29)—

1°. $\sin a : \sin b : \sin c : : \sin A : \sin B : \sin C : : \alpha : \beta : \gamma.$

$$(443)$$

2°. The first staudtian of $D - ABC$

$$= \tfrac{1}{2} \sqrt{1 - \cos^2 a - \cos^2 b - \cos^2 c + 2 \cos a \cos b \cos c}$$

$$= \sqrt{\cos a \cos b \cos c}. \tag{444}$$

3°. If M, N, P, Q, R, S be the middle points of the six edges, we have $PQ = \tfrac{1}{2} AC = \tfrac{1}{2} BD = PN$; then $PQMN$ is a lozenge, and MP is perpendicular to NQ. *Hence the three medians MP, NQ, RS form a system of three rectangular axes.*

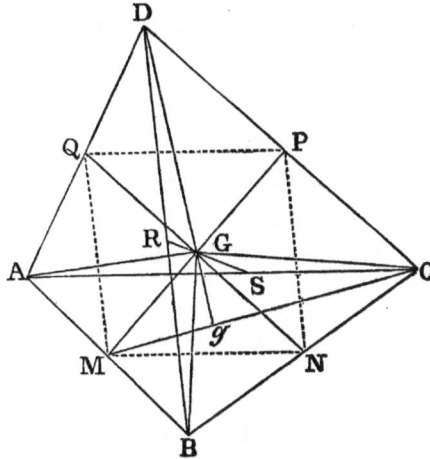

Fig. 55.

4°. Since the tetrahedron $DABC$ can coincide with $ADCB$, $BCDA$, $CBAD$, the four lines DG, AG, BG, CG are equal. *Hence G, the centre of gravity of $ABCD$, is also the centre of the circumscribed sphere, and of the inscribed sphere.*

5°. *The inscribed sphere touches the faces at the centres of the circumcircles.*

6°. *The medians PM, QN, RS are perpendicular to their corresponding edges.*

129. The four solid angles $G - ABC$, $G - ABD$, $G - BCD$, $G - ACD$, are equal. Hence the spherical triangles which they intercept on the sphere are each equal in area to one-fourth of the spherical surface, and therefore the spherical excess of each is two right angles.

Again, the angle *MGg* = supplement of *MGD* = supplement of *MGC* = *PGC* = $\frac{1}{2}AGB$. *Then, in the spherical triangle ABC, the arc which joins a summit to the middle of the opposite side is equal to the supplement of half that side, and the arc joining the point of concourse of the medians to the middle of any side is equal to half that side. Hence the spherical triangle ABC is divided by the antipodes of the point D into three diametral triangles.*

130. The volume of the tetrahedron is double of the octahedron $MNPQRS = 16SMNG = \frac{2}{3}GM \cdot GN \cdot GS$; but

$$GS^2 + GM^2 = \tfrac{1}{4}BC^2 = \tfrac{1}{4}a^2, \quad GM^2 + GN^2 = \tfrac{1}{4}\beta^2, \quad GN^2 + GS^2 = \tfrac{1}{4}\gamma^2;$$

$$\therefore \ V = \frac{\sqrt{(\beta^2 + \gamma^2 - a^2)(\gamma^2 + a^2 - \beta^2)(a^2 + \beta^2 - \gamma^2)}}{6}. \quad (445)$$

Cor. 1.—The square of the radius of the circumscribed sphere

$$= \frac{a^2 + \beta^2 + \gamma^2}{8}. \quad (446)$$

For $\quad AG^2 = AM^2 + MG^2 = \tfrac{1}{4}\gamma^2 + \dfrac{a^2 + \beta^2 - \gamma^2}{8} = \dfrac{a^2 + \beta^2 + \gamma^2}{8}.$

Cor. 2.—The radius of the inscribed sphere $= \dfrac{3V}{ABC}$. $\quad (447)$

1. If the four edges of a tetrahedron be tangents to a sphere, the sum of each pair of opposite edges is constant.

For if t_1, t_2, t_3, t_4 be the tangents drawn to the sphere from the vertices of the tetrahedron, it is evident that the sum $= t_1 + t_2 + t_3 + t_4$.

2. If a, a' be two opposite edges of a tetrahedron, and d their shortest distance, the volume

$$= \frac{1}{6}\, aa'd \sin (\overset{\wedge}{a\,a'}). \qquad (448)$$

3. The four escribed spheres of an isosceles tetrahedron are equal, and the radius of each is equal to the diameter of the inscribed sphere.

4. Prove that the radius of the sphere in Ex. 1 $= \dfrac{2t_1 t_2 t_3 t_4}{3V}$. $\qquad (449)$

5. If V be the volume of a tetrahedron, whose edges of a face are a, b, c, and opposite edges a', b', c'; and V' the volume of a tetrahedron, whose edges of a face are a', b', c', and opposite edges a, b, c; then

$$144\,(V^2 - V'^2) = (a^2 - a'^2)(b^2 - b'^2)(c^2 - c'^2). \qquad (450)$$

(Wolstenholme.)

6. If (a, a'), (b, b'), (c, c') be the three pairs of opposite edges of a tetrahedron, and denoting by the same letters the dihedral angles adjacent to these edges, prove that if the altitudes cointersect—

1°. $a^2 + a'^2 = b^2 + b'^2 = c^2 + c'^2.$ $\qquad (451)$

2°. $\cos a \cos a' = \cos b \cos b' = \cos c \cos c'.$ $\qquad (452)$

7. If the lines joining the summits of a tetrahedron to the points of contact of opposite faces with the inscribed sphere cointersect, prove that

$$\cos \tfrac{1}{2} a \cos \tfrac{1}{2} a' = \cos \tfrac{1}{2} b \,.\, \cos \tfrac{1}{2} b' = \cos \tfrac{1}{2} c \,.\, \cos \tfrac{1}{2} c'. \qquad (453)$$

8. If the spherical triangles which are equivalent to the two trihedrals $D - ABC$, $G - ABC$ (fig. Art. 127), be denoted by ABC, $A'B'C'$, respectively, prove

$$\tan \tfrac{1}{2} a' = \frac{\sin A}{\sqrt{\sin b \sin c \cos a}}. \qquad (454)$$

9. If $ABCD$ be a tetrahedron, and if we denote by $\overset{\wedge}{AB}$ the angle between the faces ABC, ABD, prove that

$$\tfrac{1}{4}\overline{AB}^2 . CD^2 . \sin^2(AB \overset{\wedge}{.} CD) = ABC^2 + ABD^2 - 2ABC . ABD \cos \overset{\wedge}{AB},$$

where ABC denotes the area of the triangle ABC. (455)

Project the triangle BCD into $B'C'D'$ on a plane perpendicular to AB; we have then

$$C'D' = CD \sin(AB \overset{\wedge}{.} CD),$$

and $$\overline{C'D'}^2 = \overline{B'C'}^2 + \overline{B'D'}^2 - 2B'C' . B'D' . \cos C'B'D';$$

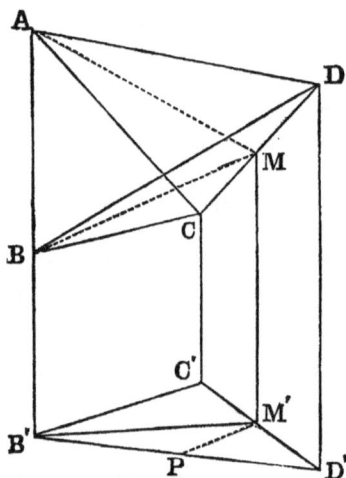

Fig. 56.

then, multiplying by AB^2, and remembering that $B'C'$, $B'D'$ are equal to the altitudes of the triangles ABC, ABD, the proposition is proved.

10. If MD be any point in the edge CD, prove that

$$ABM^2 . \overline{CD}^2 = ABC^2 . \overline{MD}^2 + ABD^2 . \overline{CM}^2 + 2ABC . ABD . CM . MD \cos \overset{\wedge}{AB}.$$

(456)

Draw $M'P$ parallel to BD, and we have

$$\overline{B'M'}^2 = B'P^2 + M'P^2 + 2B'P . PM' \cos AB;$$

also $$\frac{B'P}{B'D} = \frac{CM}{CD}, \quad \frac{PM'}{B'C'} = \frac{MD}{CD}, \quad \&c.$$

11. If M be the middle point of CD,

$$4ABM^2 = ABC^2 + ABD^2 + 2ABC . ABD . \cos \overset{\wedge}{AB}. \qquad (457)$$

CHAPTER IX.

APPLICATIONS OF SPHERICAL TRIGONOMETRY TO GEODESY AND ASTRONOMY.

SECTION I.—GEODESY.

131. *To reduce an angle to the horizon—*

1°. *General Solution.*—Let OZ be the vertical of the observer at O; then, if the angle $MON = a$, $NOZ = b$, $MOZ = c$, it is required to find the projection of MON on a horizontal plane passing through O.

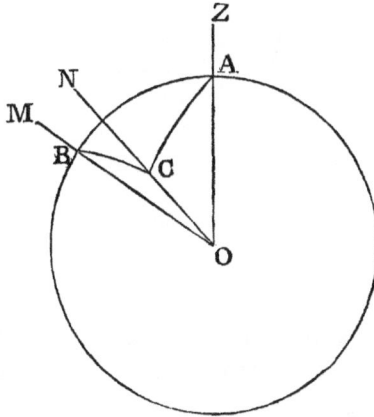

Fig. 57.

Sol.—From O as centre, with a unit radius, describe a sphere, cutting OZ, OM, ON in A, B, C, respectively; then the sought angle is the measure of the dihedral angle $BOAC$, or of the angle A of the spherical triangle BAC, which is given by the formula

$$\tan \tfrac{1}{2} A = \sqrt{\frac{\sin (s - b) \sin (s - c)}{\sin s \cdot \sin (s - a)}}.$$

2°. *Solution of Legendre.*—This solution is applicable only when the angles of elevation of the objects M, N are very small; that is, when b and c are each near $90°$. It depends on the following lemma :—*Being given a spherical triangle ABC, the angle A_1 of the rectilineal triangle formed by its chords (called the chordal triangle) is given by the equation*

$$\cos A_1 = \sin \tfrac{1}{2} b \, \sin \tfrac{1}{2} c + \cos \tfrac{1}{2} b \, \cos \tfrac{1}{2} c \, \cos A. \qquad (458)$$

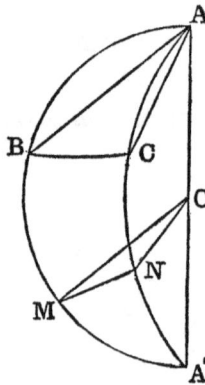

Fig. 58.

Dem.—Let $A'BC$ be the colunar triangle, M, N the middle points of the arcs $A'B$, $A'C$; then the chords AB, AC are parallel to the radii OM, ON of the sphere. Hence

$$\cos A_1 = \cos MON = \sin \tfrac{1}{2} b \, \sin \tfrac{1}{2} c + \cos \tfrac{1}{2} b \, \cos \tfrac{1}{2} c \, \cos A.$$

Cor.—If $A_1 = A - \theta$, then $\cos A_1 = \cos A + \theta \sin A$ approximately; and substituting in (458) for $\cos \tfrac{1}{2} b \, \cos \tfrac{1}{2} c$, $\sin \tfrac{1}{2} b \, \sin \tfrac{1}{2} c$, the values $\cos^2 \tfrac{1}{4} (b+c) - \sin^2 \tfrac{1}{4} (b-c)$, $\sin^2 \tfrac{1}{4} (b+c) - \sin^2 \tfrac{1}{4} (b-c)$, we get, after an easy reduction,

$$\theta = \tan \tfrac{1}{2} A \, \sin^2 \tfrac{1}{4} (b + c) - \cot \tfrac{1}{2} A \, \sin^2 \tfrac{1}{4} (b - c). \qquad (459)$$

Given the oblique angle contained between two objects above the horizon, to find the corresponding horizontal angle.

Let A be the place of the observer, M, N the objects above the horizon ; let a sphere of radius unity be described, touching the horizon at A, and intersecting the lines AM, AN in the points B, C. Draw the great circles ABO, ACO; then if H, H' denote the elevations of AM, AN above the horizon, we have $H = \frac{1}{2}$ arc AB, $H' = \frac{1}{2}$ arc AC; that is, considering the spherical triangle ABC, $H = \frac{1}{2}c$, $H' = \frac{1}{2}b$. Now the angle A of the spherical triangle ABC is the horizontal angle which corresponds to the oblique angle MAN. Hence, if θ denote the difference, we have (459)

$$\theta = \tan\tfrac{1}{2}A \, \sin^2\tfrac{1}{4}(b+c) - \cot\tfrac{1}{2}A \, \sin^2\tfrac{1}{4}(b-c);$$

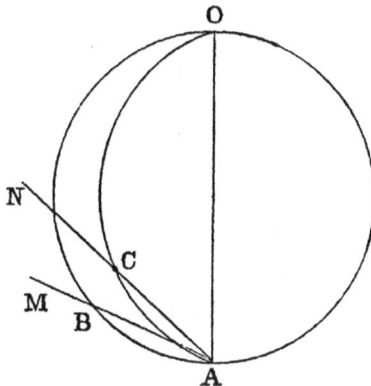

Fig. 59.

$$\therefore \; \theta = \tan\tfrac{1}{2}A \, \sin^2\tfrac{1}{2}(H+H') - \cot\tfrac{1}{2}A \, \sin^2\tfrac{1}{2}(H-H'). \quad (460)$$

In practice H, H' and θ are very small. Hence this formula may be replaced by the following, which is Legendre's :

$$\theta = \{\tfrac{1}{2}(H+H')\}^2 \tan\tfrac{1}{2} MAN - \{\tfrac{1}{2}(H-H')\}^2 \cot\tfrac{1}{2} MAN,$$
$$(461)$$

an approximate value of the difference between the circular measures of the *oblique* and *horizontal* angles, which must be added to the former to obtain the latter.

L

132. Legendre's Theorem.—

If the sides of a spherical triangle be small compared with the radius of the sphere, and if a plane triangle be constructed whose sides are equal in length to those of the spherical triangle, then each angle of the spherical triangle exceeds the corresponding angle of the plane triangle by one-third of the spherical excess.

Dᴇᴍ.—Let a, b, c be the lengths of the sides of the spherical triangle, r the radius of the sphere, then the circular measures of the sides are

$$\frac{a}{r}, \quad \frac{b}{r}, \quad \frac{c}{r},$$

respectively; hence

$$\cos A = \frac{\cos \dfrac{a}{r} - \cos \dfrac{b}{r} \cdot \cos \dfrac{c}{r}}{\sin \dfrac{b}{r} \cdot \sin \dfrac{c}{r}};$$

and, substituting for $\cos \dfrac{a}{r}$, $\cos \dfrac{b}{r}$, &c., their values given in *Pl. Trig.*, § 158, we get, neglecting powers higher than the fourth of $\dfrac{1}{r}$,

$$\cos A = \frac{\left(1 - \dfrac{a^2}{2r^2} + \dfrac{a^4}{24r^4}\right) - \left(1 - \dfrac{b^2}{2r^2} + \dfrac{b^4}{24r^4}\right)\left(1 - \dfrac{c^2}{2r^2} + \dfrac{c^4}{24r^4}\right)}{\dfrac{bc}{r^2}\left(1 - \dfrac{b^2}{6r^2}\right)\left(1 - \dfrac{c^2}{6r^2}\right)}$$

$$= \left(\frac{b^2 + c^2 - a^2}{2bc} + \frac{a^4 - b^4 - c^4 - 6b^2 c^2}{24bcr^2}\right) \div \left(1 - \frac{b^2 + c^2}{6r^2}\right)$$

$$= \left(\frac{b^2 + c^2 - a^2}{2bc} + \frac{a^4 - b^4 - c^4 - 6b^2 c^2}{24bcr^2}\right)\left(1 + \frac{b^2 + c^2}{6r^2}\right)$$

$$= \frac{b^2 + c^2 - a^2}{2bc} + \frac{a^4 + b^4 + c^4 - 2\left(a^2 b^2 + b^2 c^2 + c^2 a^2\right)}{24bcr^2}.$$

Hence, if a, β, γ denote the angles of the plane triangle, whose sides are a, b, c, we have

$$\cos A = \cos a - \frac{bc \sin^2 a}{6r^2}, \text{ nearly.}$$

Now, putting $A = a + \theta$, we have $\cos A = \cos a - \theta \sin a$.

Hence
$$\theta = \frac{bc \sin a}{6r^2} = \frac{S}{3r^2},$$

S denoting the area of the plane triangle;

$$\therefore \quad A - a = \frac{S}{3r^2}.$$

Similarly,
$$B - \beta = \frac{S}{3r^2},$$

$$C - \gamma = \frac{S}{3r^2}.$$

Hence
$$\frac{S}{r^2} = \text{spherical excess};$$

$$\therefore \quad A - a = \frac{1}{3} \text{ spherical excess.} \quad\quad (462)$$

133. The area of the spherical triangle is approximately equal to

$$S\left(1 + \frac{a^2 + b^2 + c^2}{24r^2}\right). \quad\quad (463)$$

DEM.—

$$\tan \tfrac{1}{2} E = \sqrt{\tan \frac{s}{2r} \cdot \tan \frac{s-a}{2r} \cdot \tan \frac{s-b}{2r} \cdot \tan \frac{s-c}{2r}} \, ;$$

but
$$\tan \frac{s}{2r} = \frac{\dfrac{s}{2r}\left(1 - \dfrac{s^2}{24r^2}\right)\cdots}{1 - \dfrac{s^2}{8r^2}\cdots} = \frac{s}{2r}\left(1 + \frac{s^2}{12r^2}\right).$$

Hence

$$\tan \tfrac{1}{2} E = \sqrt{\frac{s}{2r} \cdot \frac{s-a}{2r} \cdot \frac{s-b}{2r} \cdot \frac{s-c}{2r}\left(1 + \frac{s^2}{12r^2}\right)\left(1 + \frac{(s-a)^2}{12r^2}\right)\cdots} \, ;$$

therefore $\frac{1}{2}E = \frac{S}{4r^2} \sqrt{1 + \frac{s^2 + (s-a)^2 + (s-b)^2 + (s-c)^2}{12r^2}}.$

Hence $2Er^2 = S\left(1 + \frac{a^2 + b^2 + c^2}{24r^2}\right).$

Cor.—The area of the spherical triangle is equal to the area of the plane triangle, if we omit terms of the second degree in $\frac{1}{r}$.

134. If n denote the number of seconds in the spherical excess, Δ the area of the spherical triangle on the surface of the earth in square feet; then $\log n = \log \Delta - 9\cdot3267737.$—(GENERAL ROY.)

DEM.—We have $2E = \frac{n}{206265}$; $\therefore \Delta = \frac{nr^2}{206265}.$ Now the mean length of a degree $= 365155$ feet. Thus $\frac{\pi r}{180} = 365155$; substituting the value of r from this equation in the value of Δ, and taking logarithms, we get

$$\log n = \log \Delta - 9\cdot3267737. \qquad (464)$$

EXERCISES.—XXXV.

1. The angles subtended by the sides of a spherical triangle at the pole of its circumcircle are respectively double of the corresponding angles of its chordal triangle.

2. Prove Legendre's theorem from either of the formulae for $\sin \frac{1}{2}A$, $\cos \frac{1}{2}A$, $\tan \frac{1}{2}A$, respectively, in terms of the sides.

3. If the radius of the earth be 4000 miles, what is the area of a spherical triangle whose spherical excess is 1°.

4. If A'', B'', C'' be the chordal angles of the polar triangle of ABC, prove

$$\cos A'' = \sin \frac{1}{2}A \cos (s-a), \text{ &c.} \qquad (465)$$

5. If $A'BC$ be the colunar of ABC; prove that the cosines of the angles of its chordal triangle are respectively equal to

$$\cos \tfrac{1}{2}a \cos E, \quad \sin \tfrac{1}{2}b \sin (C - E), \quad \sin \tfrac{1}{2}c \sin (B - E): \qquad (466)$$

6. If R be the circumradius of a spherical triangle, A_1, B_1, C_1 the angles of its chordal triangle; prove

$$\sin A_1 = \sin \tfrac{1}{2} a \cosec R, \quad \sin B_1 = \sin \tfrac{1}{2} b \cosec R, \quad \sin C_1 = \sin \tfrac{1}{2} c \cosec R.$$

(467)

7. Prove $\sin A_1 : \sin (B_1 - C_1) :: \sin A : \sin (B - C)$. (468)

8. Prove the proposition of § 132 from equation (351).

9. Prove $E = (\tan \tfrac{1}{2} a \tan \tfrac{1}{2} b) \sin C - \tfrac{1}{2} (\tan \tfrac{1}{2} a \tan \tfrac{1}{2} b)^2 \sin 2C$. (469)
[Make use of the value of $\tan E$ drawn from equations (351), (356).]

10. Show that in every case of the solution of spherical triangles, except where the three angles are given, that Legendre's theorem may be used for an approximate solution.

Section II.—Astronomy.

135. *Astronomical Definitions.*

If *PHNR* represent the meridian of any place, produced to meet the celestial sphere, P the north pole, O the south pole of

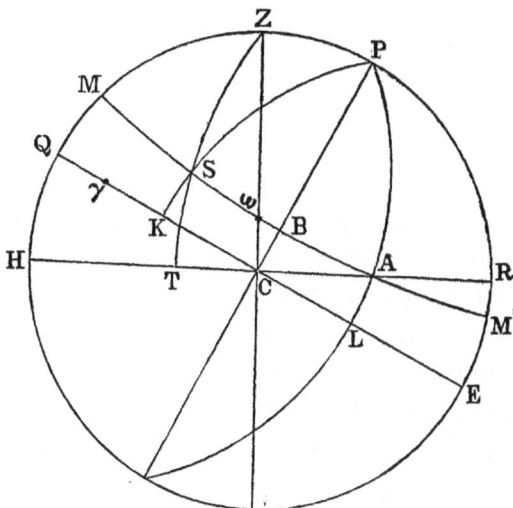

Fig. 60.

the heavens, *HR* the horizon, *EQ* the equator, Z the zenith; then, for a place whose zenith is Z, QZ is the latitude; and since *QZ* is evidently equal to *PR*, *PR* is equal to the latitude ; but *PR* is the elevation of the pole above the horizon. *Hence the elevation of the pole above the horizon is equal to the latitude.*

Again, if *S* be any heavenly body, such as the sun or a star, its position is denoted by any one of four systems of spherical co-ordinates as follows :—

1°. The great circle *ZST* passing through the zenith and *S*, and meeting the horizon in *T*, is called the *vertical circle* of *S*. The arc *HT*, measured from the south point of the horizon, or its equal the angle *HZT*, is called the *azimuth*, and *ST* the *altitude*. *HT*, *ST* are *the spherical co-ordinates of the star S* ; *ZS* is its *zenith* distance, and the arc *RT* its *azimuth from the north*.

2°. Join *SP*, and produce to meet the equator in *K*. The arcs *QK*, *KS* form the second system of spherical co-ordinates ; *QK*, or its equal the angle *ZPS*, is called the *hour angle* of *S*, and *KS* the *declination*. The declination is *positive* when *S* is north of the equator, and *negative* when south. The great circle *PSK* is called the *declination circle*, and *PS* the *polar distance* of *S*.

3°. The great circle which the centre of the sun, seen from the centre of the earth, appears to describe annually among the stars is called the *ecliptic;* and its inclination to the equator, which is nearly 23½°, *the obliquity of the ecliptic.* The points of intersection of equator and ecliptic are called the *equinoxes*— one the *vernal* equinox (called also the first point of Aries), and the other the *autumnal* equinox (the first point of Libra). If ♈ denote the first point of Aries, then ♈*K* is called the *right ascension*, and *KS* the declination of the star ; ♈*K*, *KS* are the third system of spherical co-ordinates of *S*. The right ascension is counted eastward, from 0 to 360°.

4°. From *S* draw a great circle *Sσ* perpendicular to the ecliptic ; then ♈*σ*, *σS* are the fourth system of spherical co-ordinates of *S*, and are called respectively its *longitude* and *latitude.* The longitude is reckoned eastward, from 0 to 360°. The latitude is *positive* when *north*, and *negative* when *south*.

136. If the small circle, M, M', passing through S, and parallel to the equator, represent the apparent diurnal motion of the sun or other heavenly body (the declination being supposed constant), it is evident he will be rising or setting at A (according as the eastern or the western hemisphere is represented by the diagram). He will be east or west at ω, will be at B at 6 o'clock, morning or evening, will be at noon at M, and at midnight at M'.

137. The foregoing definitions and diagram will enable us to solve several astronomical problems of an elementary character, such as the following :—

1°. *To find the time of rising or setting of a known body.*
Consider the spherical triangle APR. We have

$$\cos RPA = \tan RP \cdot \cot AP.$$

Hence, denoting the hour angle APZ by t, the latitude by ϕ, and the declination by δ, we have

$$\cos t = -\tan \phi \tan \delta. \qquad (470)$$

And the hour angle being known, the time may be found. In the case of the sun, the formula (470) gives the time from sunrise to noon, and hence the length of the day.

2°. *Being given the declination and the latitude, to find the azimuth from the north at rising.*
Let A denote the required azimuth, then $A = AR$. Hence, from the triangle ARP, we have

$$\sin \delta = \cos \phi \cdot \cos A. \qquad (471)$$

3°. *Being given the hour angle and declination of a star, to find the azimuth and altitude.*

Let Z denote the zenith distance ZS, A the azimuth from the north, p the angle ZSP at the star; then, by Delambre's Analogies,

$$\cos \tfrac{1}{2} Z . \sin \tfrac{1}{2} (p + A) = \cos \tfrac{1}{2} t . \cos \tfrac{1}{2} (\delta - \phi). \qquad (472)$$

$$\cos \tfrac{1}{2} Z . \cos \tfrac{1}{2} (p + A) = \sin \tfrac{1}{2} t . \sin \tfrac{1}{2} (\delta + \phi). \qquad (473)$$

$$\sin \tfrac{1}{2} Z . \sin \tfrac{1}{2} (p - A) = \cos \tfrac{1}{2} t . \sin \tfrac{1}{2} (\delta - \phi). \qquad (474)$$

$$\sin \tfrac{1}{2} Z . \cos \tfrac{1}{2} (p - A) = \sin \tfrac{1}{2} t . \cos \tfrac{1}{2} (\delta + \phi). \qquad (475)$$

Hence, when t, δ, ϕ are given; that is, the hour angle and declination of a heavenly body, and the latitude of the observer, z, p, A can be found. In a similar manner may be solved the converse problem :—*Given the azimuth and altitude, to find the hour angle and the declination.*

4°. If a denote the altitude of the sun at 6 o'clock, and a' the altitude when east or west; then

$$\sin a = \sin \delta . \sin \phi. \qquad (476)$$

$$\sin a' = \sin \delta \div \sin \phi. \qquad (477)$$

Exercises.—XXXVI.

1. In latitude 45° N., prove that the shadow at noon of a vertical object is three times as long when the sun's declination is 15° S. as when it is 15° N.

2. The altitude of a star when due east was 20°, and it rose *EbN*; required the latitude.

3. *Given the sun's longitude, to find his right ascension and declination.*

Fig. 61.

Let a denote the right ascension, δ the declination, ω the obliquity of the ecliptic. Now let S denote the sun's place in the ecliptic ΥS. Draw

SD perpendicular to ⋎*D*, the equator ; then, if λ denote the longitude, we have the triangle ⋎*SD*,

$$\tan \alpha = \cos \omega \tan \lambda. \qquad (478)$$

$$\sin \delta = \sin \omega \sin \lambda. \qquad (479)$$

4. Given the azimuth of the sun at setting, and at 6 o'clock, find the sun's declination, and the latitude.

5. If the sun's declination be 15° N., and length of day four hours, prove $\tan \phi = \sin 60° \tan 75°$.

6. Prove that $\qquad \cos \omega = \dfrac{\sin \phi . \tan \lambda + \sin \delta . \tan \alpha}{\sin \phi . \tan \alpha + \sin \delta . \tan \lambda}.$ $\qquad (480)$

7. Given the sun's declination and the latitude, show how to find the time when he is due east.

8. If the sun rise N.E. in latitude ϕ, prove that

$$\text{cot hour angle at sunrise} = - \sin \phi.$$

9. Given the meridian altitude and altitude when east, find the latitude and the declination.

10. Given the right ascension and the declination of a star *S*, to find its latitude and longitude.

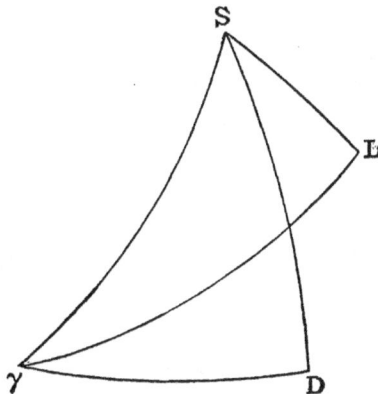

Fig. 62.

Let ⋎*D*, ⋎*L* be the equator and the ecliptic, *S* the star, *SD*, *SL* perpendicular to ⋎*D*, ⋎*L* ; then, if α be the right ascension, δ the declination, l the latitude, and λ the longitude of S, denoting the angle $S\Upsilon D$ by θ, from the right-angled triangles $S\Upsilon D$, $S\Upsilon L$, we get

$$\cot \theta = \sin \alpha \cot \delta, \quad \frac{\tan \lambda}{\tan \alpha} = \frac{\cos (\theta - \omega)}{\cos \theta}, \quad \frac{\sin l}{\sin \delta} = \frac{\sin (\theta - \omega)}{\sin \theta}. \quad (481)$$

The first of these equations determines θ, and the others λ and l.

11. Being given the latitudes and longitudes of two places on the earth considered as a perfect sphere, to find the distance between them.

This is evidently a case of § 66, viz., when two sides and the contained angle are given, to find the third side.

12. Find the latitude, being given the declination, and the interval between the time the sun is west and sunset.

13. If the latitudes and longitudes of two places on the earth be given, show how to find the highest latitude attained by a ship in sailing along a great circle from one place to the other.

14. Being given the latitudes and longitudes of two places, find the sun's declination when he is on the horizon of both at the same instant.

15. If the difference between the lengths of the longest and the shortest day at a given place be six hours, find the latitude.

16. If two stars rise together at two places, prove that the places will have the same latitude ; and if they rise together at one place, and set together at the other, the places will have equal latitudes of opposite names.

17. If ρ_1, ρ_2 be the radii vectors of two planets which revolve in circular orbits, prove, if when they appear stationary to one another, the cotangent of P_2's elongation, seen from P_1, be $\frac{1}{2} \tan \theta$, that

$$2\rho_1 = \rho_2 \tan \tfrac{1}{2} \theta \cdot \tan \theta. \quad (482)$$

18. If δ be the declination of a heavenly body, which in its diurnal motion passes in the minimum time from one to another of two parallels of altitude, whose zenith distances are Z, Z', prove that

$$\sin \delta = \frac{\cos \frac{1}{2} (Z + Z')}{\cos \frac{1}{2} (Z - Z')} \sin \text{lat}. \quad (483)$$

19. If l be the latitude, ω the obliquity of the ecliptic, prove that if the lengths of the shadow of an upright rod at noon on the longest and the shortest days be as $1 : n$,

$$\sin 2l : \sin 2\omega : : n + 1 : n - 1. \quad (484)$$

20. Determine the latitude, and the sun's declination, being given that the sun sets at 3 o'clock, and is 18° below the horizon at 4 o'clock.

21. Determine the latitudes of two places A, B from the following data :— When the sun is in the tropic of Cancer, he rises an hour earlier at A than at B; and when at the tropic of Capricorn, an hour earlier at B than at A.

22. If in latitudes l_1, l_2, l_3 on the same day, on the same meridian, the lengths of meridian shadows of towers of equal heights be s_1, s_2, s_3, prove

$$\frac{s_1 (s_2 - s_3)^2}{\tan (l_2 - l_3)} + \frac{s_2 (s_3 - s_1)^2}{\tan(l_3 - l_1)} + \frac{s_3 (s_1 - s_2)^2}{\tan (l_1 - l_2)} = 0. \qquad (485)$$

23. If the time of the sun, being due east, be midway between sunrise and 12 o'clock, find the latitude, the declination being given.

24. If the sun be due east at a given place two hours after the rising, find the declination.

25. Given the right ascension and declination of four stars, find the right ascension and declination of the point in the heavens where the diagonals of the spherical quadrilateral which they determine intersect each other.

Miscellaneous Exercises.

1. Prove that in a right-angled spherical triangle

$$\tan r = \sin (s-c), \ \tan r' = \sin (s-b), \ \tan r'' = \sin (s-a), \ \tan r''' = \sin s.$$

2. If the plane angles of a trihedral angle be respectively equal to the angles of a square, a hexagon, and a decagon, prove that the sum of its dihedral angles is five right angles.—(CATALAN.)

3. If A_1 be a chordal angle of a spherical triangle ABC, prove

$$\cos A_1 = \frac{1 + \cos a - \cos b - \cos c}{4 \sin \frac{1}{2} b \sin \frac{1}{2} c}. \tag{486}$$

4. If a spherical quadrilateral be inscribed in a small circle of the sphere, prove that the cosine of its third diagonal is equal to the product of the cosines of the tangents drawn to the small circle from the extremities of the third diagonal.

5. Prove that the volume of the pyramid whose summits are the angular points of a spherical triangle and the centre of the sphere, if the radius be equal to unity, is $\frac{1}{3}\sqrt{\tan r . \tan r_a . \tan r_b . \tan r_c}$.

6. Prove that the angles of intersection of HART's circle with the sides of a spherical triangle are $(A - B)$, $(B - C)$, $(C - A)$, respectively.

7. If in a trihedral angle $O - ABC$ we inscribe two spheres, which touch each other, if R_1, R_2 be their radii, prove that

$$\frac{R_2}{R_1} = \left\{ \sqrt{\frac{\sin (s-a) \sin (s-b) \sin (s-c)}{\sin s}} + \sqrt{\frac{1 + \sin (s-a) \sin (s-b) \sin (s-c)}{\sin s}} \right\}^3.$$

$$\text{(STEINER.)} \quad (487)$$

8. If any angle of a spherical triangle be equal to the corresponding angle of its polar triangle, prove

$$\sec^2 A + \sec^2 B + \sec^2 C + 2 \sec A \sec B \sec C \equiv 1. \tag{488}$$

9. If ABC be a diametral triangle, of which the side c is the diameter,

$$\sin^2 \tfrac{1}{2} c = \sin^2 \tfrac{1}{2} a + \sin^2 \tfrac{1}{2} b. \tag{489}$$

10. Express $\sin s$, $\sin (s - a)$, &c., in terms of the in-radii of a triangle and its colunar triangles.

11. Express $\sin E$, $\sin (A - E)$, in terms of the circumradii of a triangle and its colunars.

12. If λ, μ denote the perpendiculars from the middle point of BC on the internal and external bisectors of the angle A, prove that

$$2 \sin \lambda \sin \mu = n \cdot \sin \tfrac{1}{2} (B + C) \sec \tfrac{1}{2} a. \qquad (490)$$

13. If there be any system of fixed points A_1, A_2, A_3, &c., and a corresponding system of multiples l_1, l_2, l_3, &c., and P a point satisfying the condition $\Sigma (l \cos AP) = \text{constant}$, the locus of P is a circle.

DEM.—Let x, y, z denote the normal co-ordinates of P with respect to a fixed trirectangular triangle x_1, y_1, z_1, &c., those of A_1, &c. Then (Art. 104) we have $\Sigma (lx_1) \cdot x + \Sigma (ly_1) y + \Sigma (lz_1) z = \text{constant}$. Put $\Sigma (lx_1) = X$, $\Sigma (ly_1) = Y$, $\Sigma (lz_1) = Z$; then, if O be a point whose normal co-ordinates are

$$\frac{X}{R}, \ \frac{Y}{R}, \ \frac{Z}{R}, \ \text{where } R^2 = X^2 + Y^2 + Z^2),$$

we have $\Sigma (l \cos AP) = R \cos OP = \text{constant}$. Hence the locus of P is a circle.

Cor.—If $\Sigma (l \cos AP) = 0$, either $OP = \dfrac{\pi}{2}$, and the locus is a great circle, or $R = 0$, and then X, Y, Z must each separately vanish.

14. The sum of the cosines of the arcs, drawn from any point on the surface of a sphere to all the summits of an inscribed regular polygon, is equal to zero.

15. If O be the incentre of a spherical triangle ABC, prove that

$$\cos OA \sin (b - c) + \cos OB \sin (c - a) + \cos OC \sin (a - b) = 0. \quad (491)$$

16. If the side AB of a spherical triangle be given in position and magnitude, and the side AC in magnitude, prove, if BC meet the great circle, of which A is the pole in D, that the ratio $\cos BD : \cos CD$ is constant.

17. The eight circles tangential to any three given circles on the sphere may be divided into two tetrads, say X, Y, Z, W; X', Y', Z', W', of which one is the inverse of the other, with respect to the circle, cutting the given circles orthogonally.

18. Any three circles of either tetrad, and the non-corresponding circle of the other tetrad, are touched by a fourth circle.—(HART.)

19. Any two circles of the first tetrad, and the two corresponding circles of the second, have a fourth common tangential circle.—(*Ibid.*)

*20. If A, B, C, D be four points on the same great circle, and if ϕ be the angle of intersection of the small circles, whose spherical diameters are AC and BD, prove that the six anharmonic ratios of the points A, B, C, D are

$$\sin^2 \tfrac{1}{2}\,\phi, \quad \cos^2 \tfrac{1}{2}\,\phi, \quad -\tan^2 \tfrac{1}{2}\,\phi, \quad \operatorname{cosec}^2 \tfrac{1}{2}\,\phi, \quad \sec^2 \tfrac{1}{2}\,\phi, \quad -\cot^2 \tfrac{1}{2}\,\phi.$$

†21. The mutual power of two circles on the sphere is unaltered by inversion.

22. Prove the relation (412) by inversion.

23. If from a fixed point O on a great circle three pairs of arcs OA, OA'; OB, OB'; OC, OC' be measured, such that $\tan OA \cdot \tan OA' = \tan OB \cdot \tan OB' = \tan OC \cdot \tan OC' = k^2$, where k is a constant; then the anharmonic ratio of any four of the six points A, A', &c., which contains only one pair of conjugates, such as $(ABCC')$, is equal to the anharmonic ratio of their four conjugates $(A'B'C'C)$.—(Compare *Sequel to Euclid*, p. 132.)

Draw a tangent to the great circle, and produce the radii through the points A, A', &c., to meet the tangent.

DEF. I.—*A system of pairs of points, such as AA', BB', CC', fulfilling the conditions that the anharmonic ratio of any four being equal to that of their four conjugates, is called a system in involution.*

DEF. II.—*If two points D, D' be taken in opposite directions from O, such that $\tan^2 OD = \tan^2 OD' = k^2$, each point being evidently its own conjugate, is called a double point.*

DEF. III.—*If a system of points in involution on a great circle X be joined by arcs of great circles to any point P not on X, the six joining arcs having evidently the anharmonic ratio of the pencil formed by any four equal to that formed by their four conjugates, is called a pencil in involution.*

24. The double points D, D' are anharmonic conjugates to any pair AA' of conjugate points.

25. The six arcs joining any point on a sphere to the intersection of the sides of a spherical quadrilateral form a pencil in involution.

---

* This theorem *in plano* was first published by the author in the *Philosophical Transactions*, 1871, p. 704.

† This theorem, in a different form, viz., "the ratio of the sine squared of half the common tangent of two small circles to the product of the tangents of their radii is unaltered by inversion," was first given by the author in a Memoir "On the Equations of Circles," in the *Proceedings of the Royal Irish Academy*, 1866.

26. Any great circle is cut in involution by the sides and diagonals of a spherical quadrilateral.

27. If two diagonals of a spherical quadrilateral be quadrants, the third is a quadrant.

28. Show that the method given in "Sequel," p. 121, for describing a circle touching three circles, may be extended to the sphere.

29. Inscribe in a spherical triangle or in a small circle a triangle whose sides shall pass through three given points.

30. Prove that if CC' be the symmedian drawn from the angle C of a spherical triangle,

$$\tan CC' = 2 \sqrt{\frac{\cos^2 \tfrac{1}{2} c - \cos^2 \tfrac{1}{2} (a + b) \cos^2 \tfrac{1}{2} (a - b)}{\cot a \sin b + \cot b \sin a}}. \tag{492}$$

31. If $ABCD$ be a cyclic quadrilateral, and P any point in the circum-circle, prove that

$$\frac{\sin APB \cdot \sin CPD}{\sin APC \cdot \sin BPD} = \frac{\sin \tfrac{1}{2} AB \cdot \sin \tfrac{1}{2} CD}{\sin \tfrac{1}{2} AC \cdot \sin \tfrac{1}{2} BD}. \tag{493}$$

32. If three great circles having two points common intersect the sides of a spherical triangle in angles $\alpha_1, \alpha_2, \alpha_3$; $\beta_1, \beta_2, \beta_3$: $\gamma_1, \gamma_2, \gamma_3$, respectively, prove that

$$\begin{vmatrix} \cos \alpha_1, & \cos \alpha_2, & \cos \alpha_3 \\ \cos \beta_1, & \cos \beta_2, & \cos \beta_3 \\ \cos \gamma_1, & \cos \gamma_2, & \cos \gamma_3 \end{vmatrix} = 0. \tag{494}$$

33. Given the base of a spherical triangle and the two bisectors of the vertical angle, solve the triangle.

34. If two sides of a spherical triangle be given in position, and a point in the base fixed, if the base be bisected at the fixed point, prove that the area is either a maximum or a minimum.

35. If the sines of the perpendiculars let fall from a point on the sides of a spherical polygon, each multiplied by a given constant, be given, the locus of the point is a circle.

36. O, S are two points on the surface of the sphere; O is fixed, and S suffers a small displacement along OS proportional to $\sin OS$; prove that the displacement estimated in the directions of two great circles at right angles to each other, passing through S, are proportional to the cosines of the distances of their poles from O.

37. If a chord PQ of a small circle whose spherical centre is C pass through a fixed point O on the sphere, prove that

$$\tan \tfrac{1}{2} PCO . \tan \tfrac{1}{2} OCQ \text{ is constant.}$$

38. If through a given point O a great circle be drawn, cutting a small circle in the points A, A', and on it a point X, taken so that $\cot OX = \cot OA + \cot OA'$, the locus of X is a great circle,

39. If arcs which intersect in a point O be drawn from the angles of a triangle, meeting the opposite sides in the points A', B', C', prove

$$\frac{\tan A'O}{\tan A'O + \tan OA} + \frac{\tan B'O}{\tan B'O + \tan OB} + \frac{\tan C'O}{\tan C'O + \tan OC} = 1. \quad (495)$$

40. If a great circle, passing through a fixed point O, cut the sides of a spherical polygon in the points A, B, C, &c. ; and if X be a point, such that $\cot OX = \cot OA + \cot OB + \cot OC$, &c., the locus of X is a great circle.

INDEX.

M

Merchant Books

THE FIRST SIX BOOKS

OF THE

ELEMENTS OF EUCLID,

With Copious Annotations and Numerous Exercises.

BY

JOHN CASEY, LL.D., F.R.S.,

Fellow of the Royal University of Ireland; Member of the Council of the Royal Irish Academy; &c., &c.

Dublin: Hodges, Figgis, & Co.   London: Longmans, Green, & Co.

OPINIONS OF THE WORK.

The following are a few of the Opinions received by Dr. Casey on this Work:—

From the REV. R. TOWNSEND, F.T.C.D., &c.

"I have no doubt whatever of the general adoption of your work through all the schools of Ireland immediately, and of England also before very long."

From the "PRACTICAL TEACHER."

"The preface states that this book 'is intended to supply a want much felt by Teachers at the present day—the production of a work which, while giving the unrivalled original in all its integrity, would also contain the modern conceptions and developments of the portion of Geometry over which the elements extend.'

"The book is all, and more than all, it professes to be. . . . The propositions suggested are such as will be found to have most important applications, and the methods of proof are both simple and elegant. We know no book which, within so moderate a compass, puts the student in possession of such valuable results.

"The exercises left for solution are such as will repay patient study, and those whose solution are given in the book itself will suggest the methods by which the others are to be demonstrated. We recommend everyone who wants good exercises in Geometry to get the book, and study it for themselves."

8

ISBN: 978-1933998473

PRICE: $29.95

MB

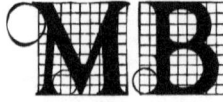